T0181712

Steel Structures Design Based on Eurocode 3

Farzad Hejazi · Tan Kar Chun

Steel Structures Design Based on Eurocode 3

Springer

Farzad Hejazi
Department of Civil Engineering
University Putra Malaysia
Selangor
Malaysia

Tan Kar Chun
Department of Civil Engineering
University Putra Malaysia
Selangor
Malaysia

Additional material to this book can be downloaded from http://extras.springer.com.

ISBN 978-981-13-4253-0 ISBN 978-981-10-8836-0 (eBook)
https://doi.org/10.1007/978-981-10-8836-0

Softcover re-print of the Hardcover 1st edition 2018

Printed on acid-free paper

This Springer imprint is published by the registered company Springer Nature Singapore Pte Ltd. part of Springer Nature
The registered company address is: 152 Beach Road, #21-01/04 Gateway East, Singapore 189721, Singapore

Preface

Steel is a better construction material compared to concrete. There are several benefits from steel construction. First of all, steel construction helps to save time. Design of steel is simpler compared to concrete. Other than that, erection of steel is faster than concrete. Steel also has post-construction advantages over concrete, in which steel can be repaired easily without affecting other members, and it can be recycled after the building is demolished.

EC3 is a design standard of steel structure, which had been enforced in the year 2010. However, in Malaysia, the usage of EC3 is still uncommon. The main reason why these phenomena had occurred is most of the designers are still not familiar with EC3. Other than that, we can barely find any guideline or reference to aid us in the design of steel structure based on EC3.

This book is tailored to the needs of structural engineers who are seeking to become familiar with the design of steel structure based on EC3.

In this book, the design procedure based on EC3 is arranged in comprehensive flowcharts. For each step, detailed explanation and all the necessary table/equation will be provided. Other than that, examples also provided to show the proper way to perform design. This book also provides useful appendix, including universal sections and their properties, and general formula of shear force, maximum bending moment, and deflection for several selected loading condition. These appendices serve to give convenience to the designers when they are performing design.

This book also introduces a specially developed design-aiding program. This program can give the immediate result to the user after it receives inputs from the user. With this program, modeling is not required and the time consumed in design stage can be reduced.

Selangor, Malaysia

Farzad Hejazi
Tan Kar Chun

Contents

Chapter 1
Introduction

1.1 General

Steel is a material commonly used in construction. In concrete structures, steel is mainly used as reinforcement to increase the resistance of the concrete member in the tension zone. In steel structures, steel is important because the structural members are constructed purely from structural steel.

Steel is an alloy of iron and carbon, with carbon contributing between 0.2 and 2% of the weight of steel. If the alloy contains less than 0.2% carbon, it is called wrought iron, which is soft and malleable. If the alloy contains more than 3% carbon, it is called cast iron, which is hard and brittle.

Structural steel is basically carbon steel, which is steel with controlled amounts of manganese, phosphorus, silicon, sulfur, and oxygen added. Carbon steel can be further categorized according to its carbon content: mild steel (0.2–0.25% carbon), medium steel (0.25%–0.45%), hard steel (0.45–0.85%), and spring steel (0.85–1.85%).

As steel is a construction material, designers must know its mechanical properties. The notable mechanical properties of steel are as follows:

- Modulus of elasticity, $E = 210 \times 10^9$ N/m^2
- Shear modulus, $G = 81 \times 10^9$ N/m^2
- Poisson's ratio, $v = 0.3$

F. Hejazi and T. K. Chun, *Steel Structures Design Based on Eurocode 3*,
https://doi.org/10.1007/978-981-10-8836-0_1

1.2 Advantages of Steel Structure

Figure 1.1 shows some of the advantages of steel over reinforced concrete in construction. The design of a steel structure is simpler than that of a concrete structure. In the design of a concrete structure, factors such as member dimension, diameter of steel bar, and concrete grade must be determined, all of which lead to uncertainty and variations in the design outcome. By contrast, the design of a steel structure is fundamentally based on standard sections, which reduces uncertainty and variations in the design outcome.

Another advantage of steel over concrete is that it can be constructed under all kinds of weather. Given that steel frames can be fabricated off-site, the effect of weather on the progress of the project is minimal. On the contrary, concrete frames are commonly cast on-site, where bad weather conditions can hinder the progress of the project.

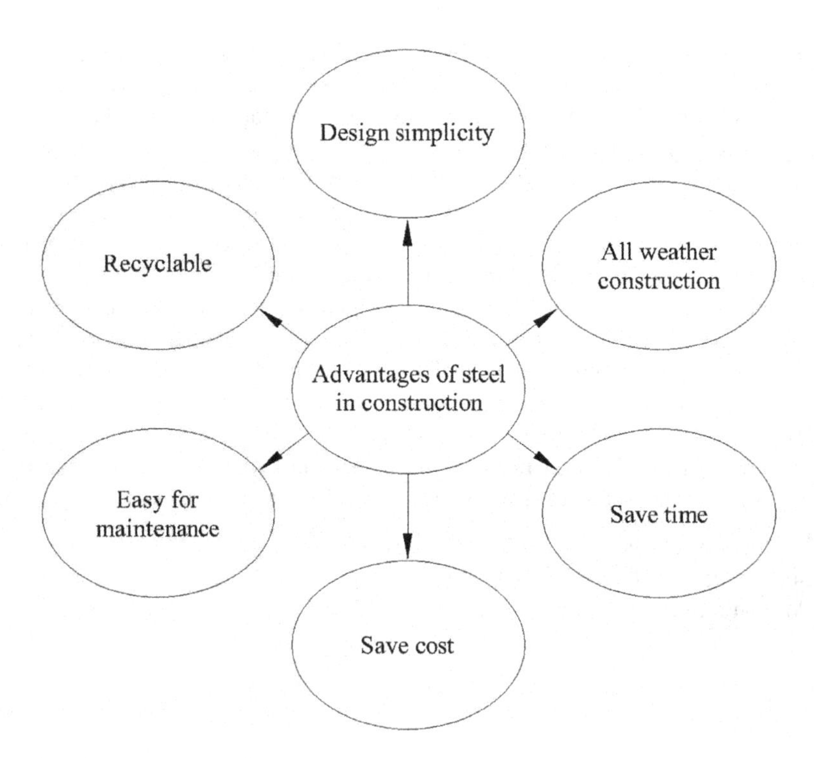

Fig. 1.1 Advantages of steel in construction

The construction of a steel structure is also easy because it only employs the welding or bolting process. Therefore, construction can be finished in a short time. Fabrication of concrete, however, takes a long time because of the casting and curing process involved.

Both all-weather construction and ease of construction can efficiently reduce project duration, which is favorable for owners because they can generate profit as early as possible.

1.3 Design Standard for Steel

Eurocode 3 (EC3) is a design standard belonging to a set of harmonized technical rules called Eurocodes. Eurocodes were developed by the European Committee of Standardization to remove all design obstacles and harmonize technical specifications in European countries. In 2010, the previously implemented BS 5950 was superseded by EC3. The change in design standard was claimed to improve the construction industry because EC3 allows for a more economical design compared with BS 5950. In addition, the newly established EC3 is well arranged, less restrictive, and more logical compared with its predecessor.

The design under Eurocodes is based on a limit state. Limit-state designs have two types: ultimate limit state (ULS) and serviceability limit state (SLS).

ULS design is concerned with structural stability under the ultimate condition, whereas SLS design is concerned with structural function under normal use, occupant comfort, and building appearance. ULS and SLS designs can be carried out by applying different partial safety factors to a load, as shown in Table 1.1.

During the design stage, one of the most important tasks, and also the most difficult, is estimating the load to be applied to a structure. In design, load can be classified as dead load (DL) and live load (LL).

Table 1.1 Load combinations for ULS and SLS designs (BS EN 1990 Table NA.A1.2)

Load combination for ultimate limit state design	Load combination for serviceability limit state design
$1.35G_k + 1.5Q_k$	$G_k + Q_k$
$1.35G_k + 1.5W_k$	$G_k + W_k$
$1.00G_k + 1.5W_k$	$G_k + Q_k + 0.5W_k$
$1.35G_k + 1.5Q_k + 0.75W_k$	$G_k + Q_k + W_k$
$1.35G_k + 1.05Q_k + 1.5W_k$	

DL is defined as a permanent action (G_k) in Eurocodes, that is, the load permanently attached to a structure. Therefore, it is basically the self-weight of a material for either structural or architectural purposes.

LL is defined as a variable action (Q_k) in Eurocodes, that is, the load induced from activities. It is mostly induced from human activities for most structures. In a bridge, for instance, traffic load is considered instead. In Eurocodes, the design values of LLs at different locations are provided.

Wind load (WL) is a type of LL. It is usually not considered except for tall buildings. This load is hugely dependent on the terrain and location where the building stands and the building height. Design values for WL can be obtained from the national standard instead of from Eurocodes.

After the load is estimated, the next step is to determine the load combination. Table 1.1 shows several options for load combinations for ULS and SLS.

1.4 I-Section

One of the most commonly used steel member sections is the I-section, also known as the universal section. Figure 1.2 shows the terminology and dimensions of an I-section.

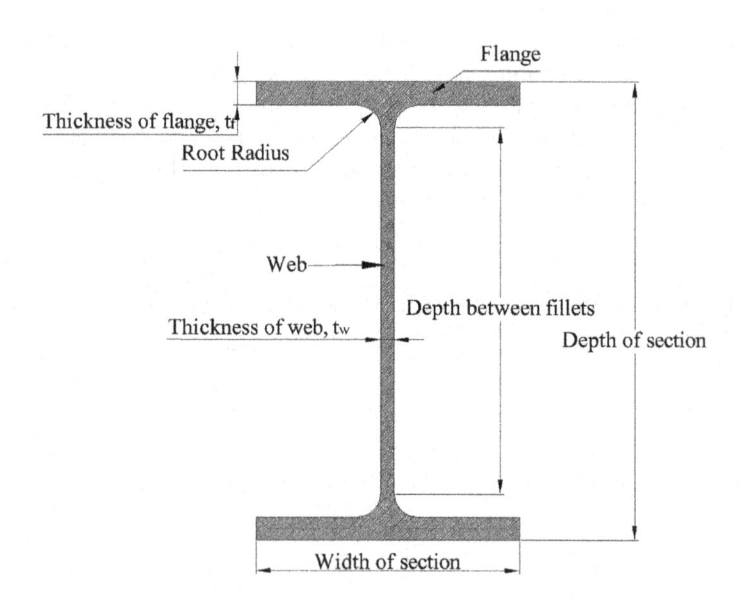

Fig. 1.2 Terminology and dimension of an I-section

1.5 Steel Design Based on EC3 Program

An special program is developed for "Steel Design Based on EC3". The program can perform three types of design, which is design of beam, column and connection (Fig. 1.3).

This is a simple complementary program that gives quick result for design of beam, column and connection.

The program can be downloaded through the following link: http://extras.springer.com

In the main menu, one of the following options can be choose: "Design of Beam", "Design of Column (Simple Construction)" or "Design of Connection", and then click START to proceed.

For "Design of Beam" and "Design of Column (Simple Construction)", select the section to use before proceed to design.

- In order to design a beam, the structural analysis is required. By specifying the supports condition and length, the structural loading will be calculated. Then, this result will be used as design input, which will yield the section to use at the end.
- To design a column, column support condition, length and loading on each direction is required. Similarly, the program will determine the optimum section for the loading condition.
- Design of connection included bolted connection and welded connection. For bolted connection, parameter for components involved in construction of

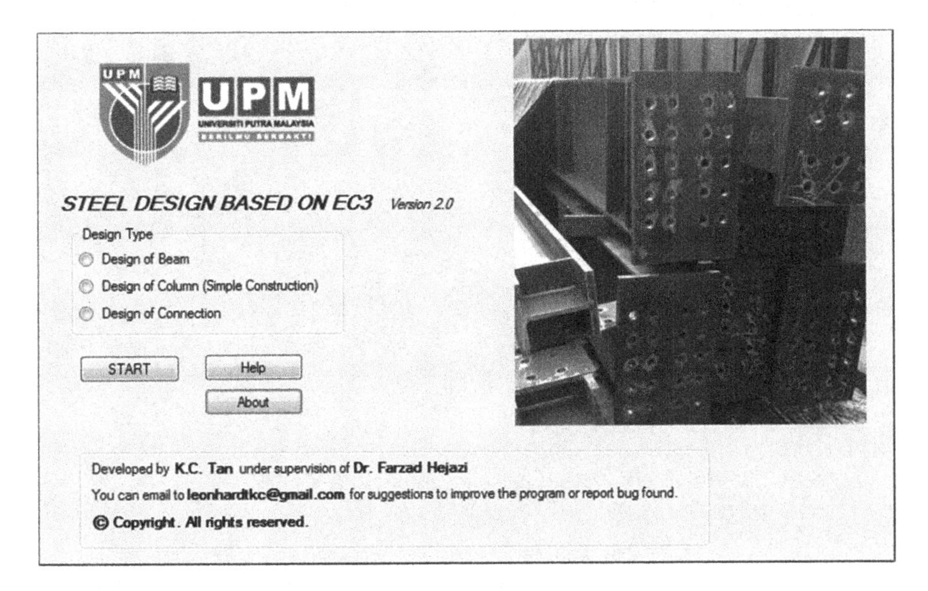

Fig. 1.3 Main menu of steel design based on EC3 program

connection such as steel plate and bolt, as well as design load is required. The program will determine the number of bolt required for the considered condition. For welded connection, the steel plate parameter and design load are required as input, while the program will determine the welding length required for the considered condition.

The result generated from the program can be exported to Microsoft Excel worksheet format. The output file of the program can be implemented as design outcome.

Chapter 2
Beam Design

2.1 Introduction

Beam is a structural member subjected to a transverse load, whose direction is perpendicular to the longitudinal axis (x-x) of the beam. Thus, a beam is designed to resist the bending moment and shear force of the load. Generally, a beam is bent about its major axis (y-y) (Fig. 2.1).

Beams can be categorized into two types: primary and secondary. A primary beam supports a secondary beam and a slab while being supported only by a column. A secondary beam supports a slab while being supported by a primary beam or a column. Steel beams can also be categorized as laterally restrained and laterally unrestrained. Lateral rotation and deflection are not allowed for a laterally restrained beam. Figure 2.2 shows examples of laterally restrained beams.

By contrast, a laterally unrestrained beam is free to rotate and deflect laterally when load is applied. Any beam without restraints on its sides is categorized as a laterally unrestrained beam.

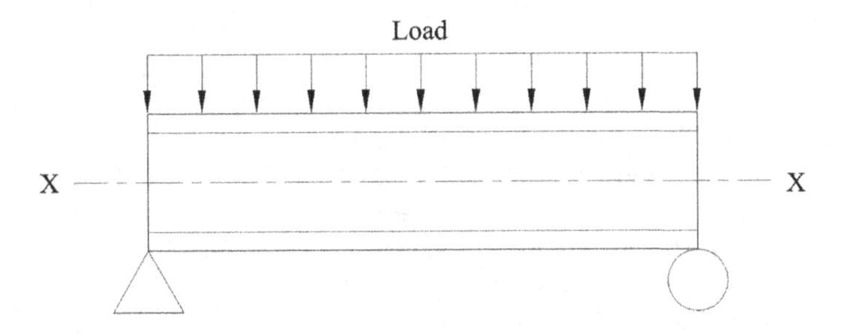

Fig. 2.1 Beam and its loading

F. Hejazi and T. K. Chun, *Steel Structures Design Based on Eurocode 3*,
https://doi.org/10.1007/978-981-10-8836-0_2

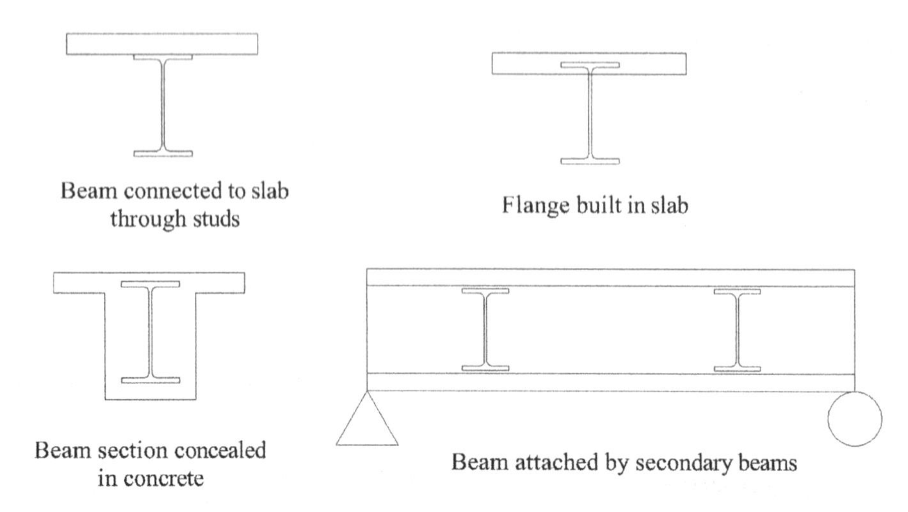

Fig. 2.2 Examples of laterally restrained beams

Table 2.1 Nominal values of yield strength f_y and ultimate tensile strength f_u for hot-rolled structural steel (BS EN 1993-1-1:2005 Table 3.1)

Standard and steel grade (To BS EN 10025-2)	Nominal thickness of element, t (mm)			
	$t \leq 40$ mm		$40 \text{ mm} < t \leq 80$ mm	
	f_y (N/mm^2)	f_u (N/mm^2)	f_y (N/mm^2)	f_u (N/mm^2)
S235	235	360	215	360
S275	275	430	255	410
S355	355	490	335	470
S450	440	550	410	550

2.2 Design Procedure for a Laterally Restrained Beam

The design procedure for a laterally restrained beam is presented below.

1. Determine the support condition (i.e., pin, roller, or fixed at both ends of the beam).
2. Determine the DL and LL that act on the beam.
3. Choose the steel grade. Refer to BS 4 Part 1 2005 to choose the beam section for use in construction. A table for the universal beam section and its corresponding properties is provided in Appendix A.2 (Table 2.1).
4. Perform a structural analysis to determine the maximum shear force V_{Ed} and bending moment M_{Ed} induced by loading. Prior to analysis, the partial safety factor for ULS (Table 1.1) is applied to the actions determined in Step 2, including the self-weight of the beam section.

5. Classify the beam section. For beams, check only the section class by using the criteria "outstand flange for rolled sections" and "web with neutral axis at mid-depth, rolled sections" (Table 2.2).
6. Determine shear resistance of the section. The shear area of the section needs to be determined beforehand. γ_{MO} should be set as 1.0.

$$V_{pl,Rd} = \frac{A_V \left(f_y/\sqrt{3}\right)}{\gamma_{MO}} \quad (2.1)$$

where

A_V is shear area obtained from Step 6 (Table 2.3)
f_y is yield strength of steel obtained from Step 3

(BS EN 1993-1-1:2005 6.2.6(2))

7. Compare the design shear force on the structure and shear resistance of the section. If the shear resistance of the structure is insufficient, repeat Step 3 to choose a better section. Otherwise, proceed to Step 8.

Table 2.2 Maximum width-to-thickness ratio of the compression element (BS EN 1993-1-1:2005 Table 5.2)

Type of element	Class of element		
	Class 1	Class 2	Class 3
Outstand flange for rolled section	$c/t_f \leq 9\varepsilon$	$c/t_f \leq 10\varepsilon$	$c/t_f \leq 14\varepsilon$
Web with neutral axis at mid depth, rolled sections	$c^*/t_w \leq 72\varepsilon$	$c^*/t_w \leq 83\varepsilon$	$c^*/t_w \leq 124\varepsilon$
Web subject to compression, rolled sections	$c^*/t_w \leq 33\varepsilon$	$c^*/t_w \leq 38\varepsilon$	$c^*/t_w \leq 42\varepsilon$
f_y	235	275	355
ε	1	0.92	0.81

Where t_f is thickness of flange by referring to Appendix A.2
t_w is thickness of web by referring to Appendix A.2
$c^* = d$ by referring to Appendix A.2
$c = (b - t_w - 2r)/2$

Table 2.3 Shear area, A_V, parameter descriptions (BS EN 1993-1-1:2005 6.2.6(3))

Type of member	Shear area, A_V
Rolled I and H sections, load parallel to web	$A - 2bt_f + (t_w + 2r)t_f \geq \eta h_w t_w$
Rolled channel sections, load parallel to web	$A - 2bt_f + (t_w + r)t_f$
Rolled rectangular hollow sections of uniform thickness, load parallel to depth	$Ah/(b + h)$
Circular hollow sections and tubes of uniform thickness	$2A/\pi$
Plates and solid bars	A

8. Check whether the section is classified as a plated member. This step is especially necessary for a built-up section because universal beam sections usually do not satisfy Eq. 2.2, in which case, Step 9 is skipped. Otherwise, the shear buckling resistance of the section should be determined according to BS EN 1993-1-5. η is set as 1.0.

$$\frac{h_w}{t_w} > 72\frac{\varepsilon}{\eta} \tag{2.2}$$

where $h_w = d + 2r$

d is depth between fillets by referring to Appendix A.2
r is root radius by referring to Appendix A.2
t_w is thickness of web by referring to Appendix A.2
ε is obtained from Step 5 (Table 2.2)

(BS EN 1993-1-1:2005 6.2.6(6))

9. Determine the shear buckling resistance according to BS EN 1993-1-5.
10. Determine the bending moment resistance of the section. Note that for a different section class, the section properties used are different.

$$M_{C,Rd}\begin{cases}\frac{W_{pl}f_y}{\gamma_{M0}}, \textit{Class } 1 \textit{ and } 2 \textit{ sections}\\ \frac{W_{el,min}f_y}{\gamma_{M0}}, \textit{Class } 3 \textit{ sections}\\ \frac{W_{eff,min}f_y}{\gamma_{M0}}, \textit{Class } 4 \textit{ sections}\end{cases} \tag{2.3}$$

where

W_{pl} is plastic section modulus by referring to Appendix A.2
$W_{el,min}$ is minimum elastic section modulus
$W_{eff,min}$ is minimum effective section modulus
f_y is yield strength of steel obtained from Step 3 (Table 2.1)

(BS EN 1993-1-1:2005 6.2.5(2))

11. Compare the design bending moment of the structure and the bending moment resistance of the section. If the bending moment resistance of the structure is insufficient, repeat Step 3 to choose a better section. Otherwise, proceed to Step 12.
12. Refer to BS EN 1993-1-1:2005 6.2.8(2) to check the ratio of design shear force to shear resistance of the section. If the ratio is more than 0.5, proceed to Step 13. Otherwise, proceed to Step 15 to continue with the design.
13. Determine the reduced bending moment resulting from the shear force. The formula for bending moment resistance remains unchanged, as shown in Eq. 2.3, but the value of f_y is replaced by f_{yr}. Alternatively, reduced bending moment can be determined directly if the section has equal flanges.

$$f_{yr} = (1 - \rho)f_y$$

$$\rho = \begin{cases} \left(\frac{2V_{Ed}}{V_{pl,Rd}} - 1\right)^2, generally \\ \left(\frac{2V_{Ed}}{V_{pl,T,Rd}} - 1\right)^2, with\ Torsion \end{cases}$$

Alternatively,

$$M_{y,Rd} = \frac{\left(W_{pl,y} - \frac{\rho A_w^2}{4t_w}\right)f_y}{\gamma_{M0}} \tag{2.4}$$

where

V_{Ed} is design shear force obtained from Step 4
$V_{pl,Rd}$ is design shear resistance obtained from Step 6 (Eq. 2.1)
$V_{pl,T,Rd}$ is design shear resistance that take torsion into account
f_y is yield strength of steel obtained from Step 3 (Table 2.1)
$W_{pl,y}$ is plastic section modulus by referring to Appendix A.2
t_w is thickness of web by referring to Appendix A.2

$$A_w = h_w t_w;\ h_w = d + 2r$$

d is depth between fillets by referring to Appendix A.2
r is root radius by referring to Appendix A.2

(BS EN 1993-1-1:2005 6.2.8(3), (4), (5))

14. Compare the design bending moment of the structure and the reduced bending moment resistance of the section. If the bending moment resistance of the structure is insufficient, repeat Step 3 to choose a better section. Otherwise, proceed to Step 15.
15. Determine the maximum deflection of the structure under the loading specified in Step 2. The load combination for this calculation should be any of those specified for the SLS design, as shown in Table 1.1.
16. Determine the allowable deflection of the structure (Table 2.4).

Table 2.4 Vertical deflection limit Δ_{all} (BS EN 1993-1-1:2005 NA2.23)

Design situation	Vertical deflection limit, Δ_{all}
Cantilever	$Length/180$
Beams carrying plaster of other brittle finish	$Length/360$
Other beams (except purlins and sheeting rails)	$Length/200$
Purlins and sheeting rails	To suit the characteristics of particular cladding

17. Compare the maximum deflection of the structure and the allowable deflection. If the deflection of the structure exceeds the allowable deflection, repeat Step 3 to choose a better section. Otherwise, proceed to Step 18.
18. Check whether the section is an overdesign by checking the ratio of design value to resistance for shear and bending and the ratio of maximum deflection to allowable deflection. If both ratios are less than 0.5, repeat Step 3 and choose a smaller section to ensure optimum design.

2.2.1 Design Flowchart for a Laterally Restrained Beam

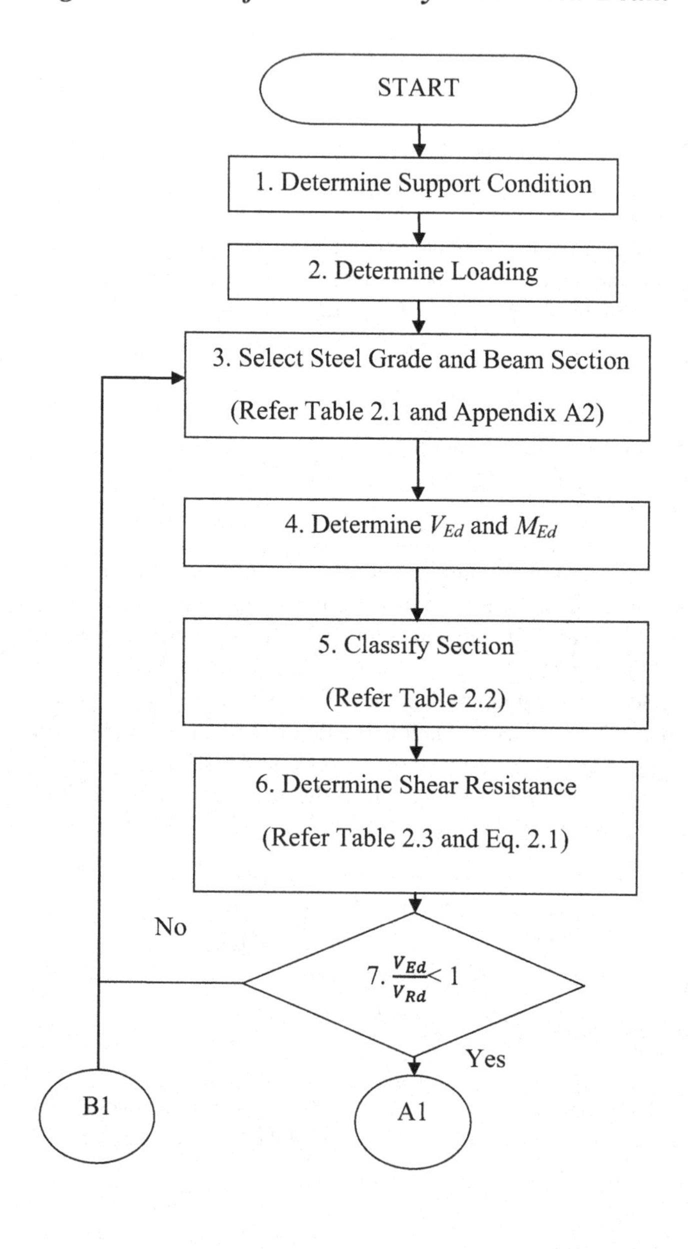

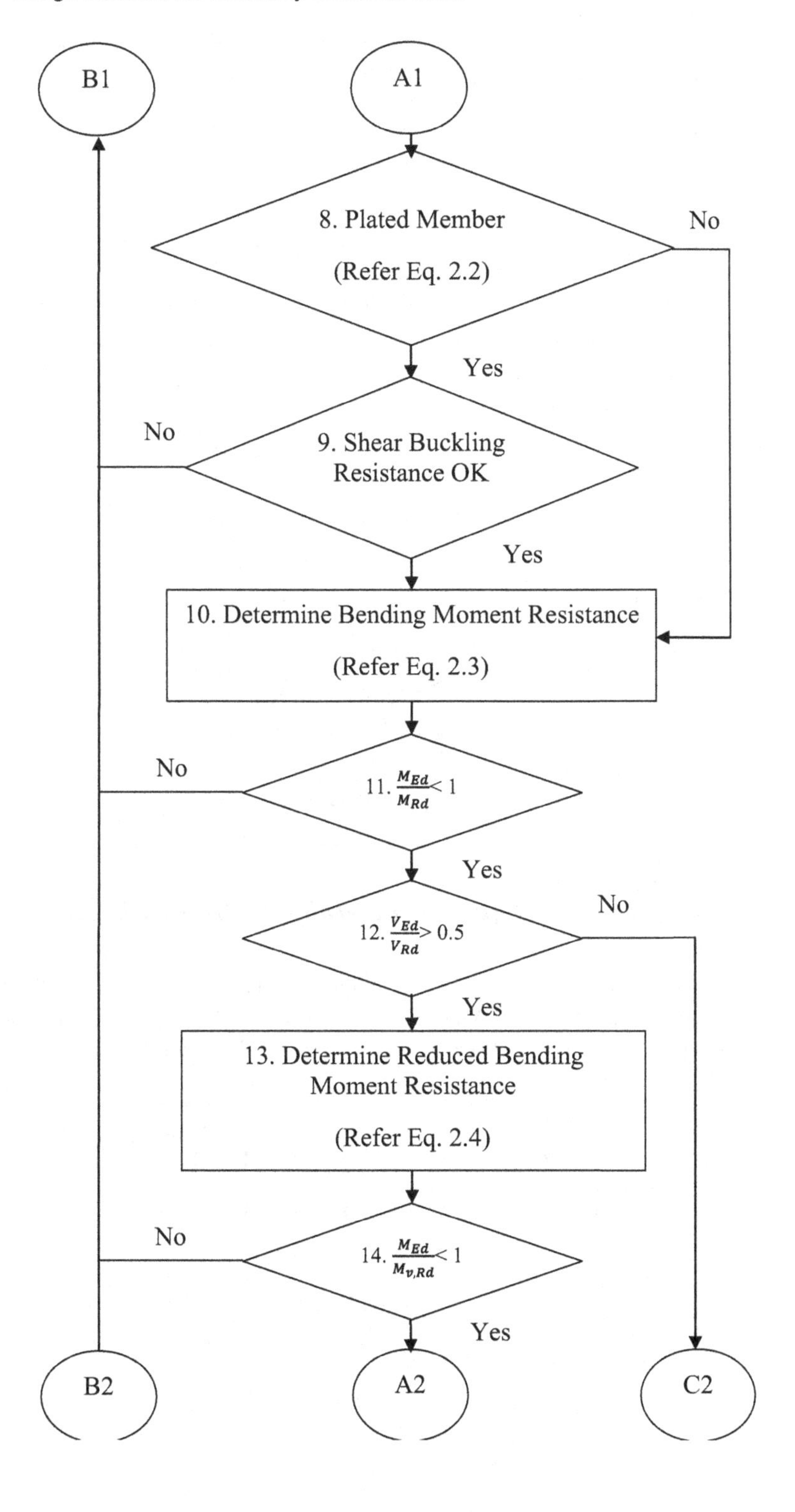
B1
A1
8. Plated Member
(Refer Eq. 2.2)
No
Yes
No
9. Shear Buckling Resistance OK
Yes
10. Determine Bending Moment Resistance
(Refer Eq. 2.3)
No
11. $\frac{M_{Ed}}{M_{Rd}} < 1$
Yes
No
12. $\frac{V_{Ed}}{V_{Rd}} > 0.5$
Yes
13. Determine Reduced Bending Moment Resistance
(Refer Eq. 2.4)
No
14. $\frac{M_{Ed}}{M_{v,Rd}} < 1$
Yes
B2
A2
C2

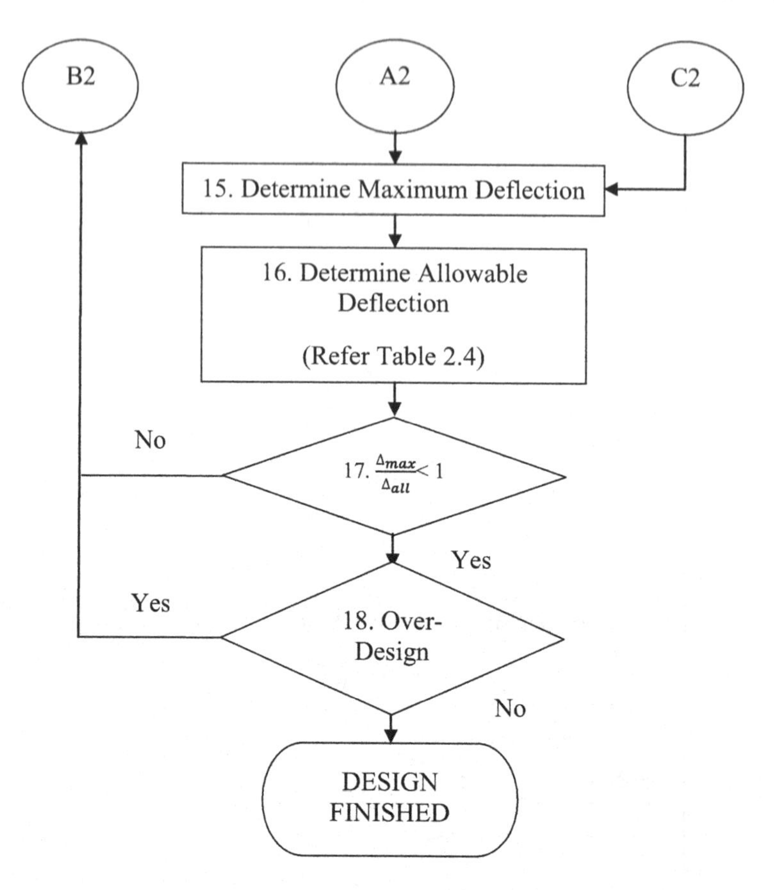

2.2.2 *Example 2-1 Design of a Laterally Restrained Beam*

Select the optimum section of a beam 5 m in length and subjected to a uniform load (Fig. 2.3). Use steel grade S235. Assume the beam is laterally restrained and sits on 100 mm bearings at each end. Take the self-weight of the beam into account.

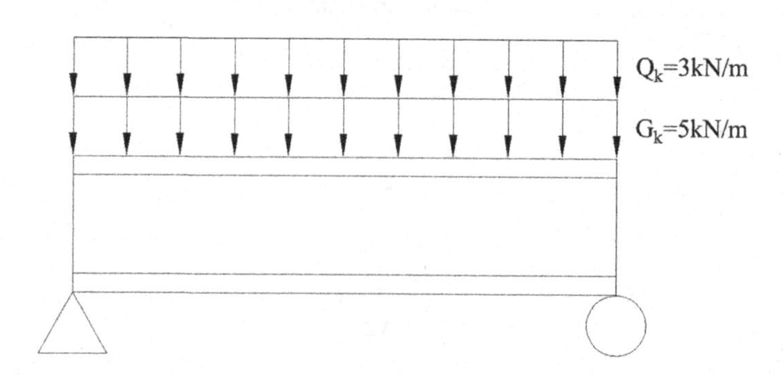

Fig. 2.3 Example 2-1

Step	Reference	Action/calculation	Conclusion
1	References are to BS EN 1993-1-1 unless otherwise stated	The beam is **simply supported**	
2		Permanent action, G_k = **5 kN/m** Variable action, Q_k = **3 kN/m**	
3	Table 3.1	Steel grade = **S235** Assume the thicknesses of web and flange are less than 40 mm: f_y = **235 N/mm²**	$f_y = 235$ N/mm²
	BS 4 Part 1 2005	Randomly choose a beam section for the first trial: Select beam section **305 × 127 × 37** The properties of the section is as follows: Mass per meter = 37 kg/m Depth of section, D = 304.4 mm Width of section, b = 123.4 mm Thickness of web, t_w = 7.1 mm Thickness of flange, t_f = 10.7 mm Root radius, r = 8.9 mm Depth between fillets, d = 265.2 mm Second moment of area about major (y-y) axis, I_y = 7171 cm⁴ Elastic modulus about major (y-y) axis, $W_{el,y}$ = 471 cm³ Plastic modulus about major (y-y) axis, $W_{pl,y}$ = 539 cm³ Area of section, A = 47.2 cm²	
4		Self-weight of beam section = 37 kg/m × 9.81 N/kg = **0.36 kN/m** For ULS, partial factor of safety for both permanent action and variable action selected are 1.35 and 1.5 respectively Ultimate load, w_{ult} $= 1.35G_k + 1.5Q_k$ = 1.35(5 + 0.36) + 1.5(3) = **11.74 kN/m**	Design load = 11.74 kN/m
		For simply supported beam, V_{Ed} and M_{Ed} can be determined using equation below: V_{Ed} $= \frac{w_{ult}L}{2}$ $= \frac{11.74 \times 5}{2}$ = **29.35 kN**	V_{Ed} = 29.35 kN
		M_{Ed} $= \frac{w_{ult}L^2}{8}$ $= \frac{11.74 \times 5^2}{8}$ = **36.69 kNm**	M_{Ed} = 36.69 kNm

(continued)

(continued)

Step	Reference	Action/calculation	Conclusion
5	Table 5.2	Section classification: i. f_y = 235 N/mm^2 $\varepsilon = 1$ **Class 1** ii. Rolled section, outstand flange: $c = \frac{b-t_w-2r}{2}$ $= \frac{123.4-7.1-2(8.9)}{2}$ = 49.25 mm t_f = 10.7 mm $\frac{c}{t_f} = \frac{49.25}{10.7} = 4.60 < 9\epsilon (= 9)$ **Class 1** iii. Rolled section, web with neutral axis at mid depth: $c^* = d$ = 265.2 mm t_w = 7.1 mm $\frac{c^*}{t_w} = \frac{265.2}{7.1} = 37.35 < 72\epsilon (= 72)$ **Class 1** Therefore, the section is **class 1**	Section class 1
6	6.2.6(3)	For I beam with load applied on flange, consider the case of rolled I sections with load parallel to web: Shear area, A_v $= A - 2bt_f + (t_w + 2r)t_f$ $= 47.2 \times 10^2 - 2(123.4)$ $(10.7) + (7.1 + 2(8.9))(10.7)$ = 2345.67 mm^2	
	6.2.6(2)	$V_{pl,Rd} = \frac{A_v(f_y/\sqrt{3})}{\gamma_{M0}}$ $= \frac{2345.67 \times 235}{\sqrt{3}}$ = **318.25 kN**	$V_{pl,Rd}$ = 318.25 kN
7		$\frac{V_{Ed}}{V_{pl,Rd}} = \frac{29.35}{318.25} = 0.09 < 1$ The shear resistance of the section is adequate	$\frac{V_{Ed}}{V_{pl,Rd}} = 0.09$
8	6.2.6(6)	Check for shear buckling failure: $h_w = d + 2r$ = 265.2 + 2(8.9) = 283 mm t_w = 7.1 mm $\frac{h_w}{t_w} = \frac{283}{7.1} = 39.86 < 72\frac{\epsilon}{\eta} (= 72)$ Shear buckling check is not required	
9		*This step is skipped as shear buckling check is not required*	
10	6.2.5(2)	For Class 1 section, Bending moment resistance, $M_{c,Rd} = M_{pl,Rd}$ $= \frac{W_{pl}f_y}{\gamma_{M0}}$ $= \frac{539 \times 10^{-6} \times 235 \times 10^6}{1}$ = **126.67 kNm**	$M_{c,Rd}$ = 126.67 kNm

(continued)

(continued)

Step	Reference	Action/calculation	Conclusion
11		$\frac{M_{Ed}}{M_{c,Rd}} = \frac{36.69}{126.67} = \mathbf{0.29} < 1$ The bending resistance of the section is adequate	$\frac{M_{Ed}}{M_{c,Rd}} = 0.29$
12	6.2.8(2)	Check for combination of shear and bending failure: $\frac{V_{Ed}}{V_{pl,Rd}} = \frac{29.35}{318.25} = \mathbf{0.09} < 0.5$ Reduction in bending resistance is not required	
13		*This step is skipped as reduction in bending resistance is not required*	
14		*This step is skipped as reduction in bending resistance is not required*	
15		For SLS, partial factor of safety for both permanent action and variable action selected is 1.0. Serviceability load, w_{ser} $= 1.0G_k + 1.0Q_k$ $= 1.0(5.36) + 1.0(3)$ $= 8.36$ kN/m For simply supported beam, maximum deflection can be determined using equation below: Maximum deflection, Δ_{max} $= \frac{5wL^4}{384EI}$ $= \frac{5 \times 8.36 \times 10^3 \times 5^4}{384 \times 210 \times 10^9 \times 7171 \times 10^{-8}}$ $= 4.52 \times 10^{-3}$ m $=$ **4.52 mm**	$\Delta_{max} = 4.52$ mm
16	NA2.23	Assume the beam carries plaster of other brittle finishes: Allowable deflection, Δ_{all} $= \frac{L}{360}$ $= \frac{5}{360}$ $= 0.01389$ m $=$ **13.89 mm**	$\Delta_{all} = 13.89$ mm
17		$\frac{\Delta_{max}}{\Delta_{all}} = \frac{4.52}{13.89} = \mathbf{0.33} < 1$ The deflection is allowable	$\frac{\Delta_{max}}{\Delta_{all}} = 0.33$
18		Check the following ratio: $\frac{V_{Ed}}{V_{pl,Rd}} = \frac{29.35}{318.25} = \mathbf{0.09}$ $\frac{M_{Ed}}{M_{c,Rd}} = \frac{36.69}{126.67} = \mathbf{0.29}$ $\frac{\Delta_{max}}{\Delta_{all}} = \frac{4.52}{13.89} = \mathbf{0.33}$ All ratios are significantly small. Therefore, the beam section 305 × 127 × 37 is **not optimum**	

Step 3 is repeated in the design process because the optimum section is required (Fig. 2.4).

Step	Reference	Action/calculation	Conclusion
3		Steel grade = **S235** Assume the thicknesses of web and flange are less than 40 mm: $\mathbf{f_y = 235\ N/mm^2}$	f_y = 235 N/mm^2
	BS 4 Part 1 2005	Select beam section **254** × **102** × **22** The properties of the section is as follows: Mass per meter = 22 kg/m Depth of section, D = 254.0 mm Width of section, b = 101.6 mm Thickness of web, t_w = 5.7 mm Thickness of flange, t_f = 6.8 mm Root radius, r = 7.6 mm Depth between fillets, d = 225.2 mm Second moment of area about major (y-y) axis, I_y = 2841 cm^4 Elastic modulus about major (y-y) axis, $W_{el,y}$ = 224 cm^3 Plastic modulus about major (y-y) axis, $W_{pl,y}$ = 259 cm^3 Area of section, A = 28.0 cm^2	
4		Self-weight of beam section = 22 kg/m × 9.81 N/kg = **0.22 kN/m**	
		For ULS, partial factor of safety for both permanent action and variable action selected are 1.35 and 1.5 respectively. Ultimate load, w_{ult} $= 1.35G_k + 1.5Q_k$ $= 1.35(5 + 0.22) + 1.5(3)$ = **11.55 kN/m**	Design load = 11.55 kN/m
		For simply supported beam, V_{Ed} and M_{Ed} can be determined using equation below: V_{Ed} $= \frac{w_{ult}L}{2}$ $= \frac{11.55 \times 5}{2}$ = **28.88 kN**	V_{Ed} = 28.88 kN
		M_{Ed} $= \frac{w_{ult}L^2}{8}$ $= \frac{11.55 \times 5^2}{8}$ = **36.09 kNm**	M_{Ed} = 36.09 kNm

(continued)

(continued)

Step	Reference	Action/calculation	Conclusion
5	Table 5.2	Section classification: i. $f_y = 235$ N/mm^2 $\varepsilon = 1$ **Class 1** ii. Rolled section, outstand flange: $c = \frac{b-t_w-2r}{2}$ $= \frac{101.6-5.7-2(7.6)}{2}$ $= 40.35$ mm $t_f = 6.8$ mm $\frac{c}{t_f} = \frac{40.35}{6.8} = 5.93 < 9\epsilon(= 9)$ **Class 1** iii. Rolled section, web with neutral axis at mid depth: $c^* = d$ $= 225.2$ mm $t_w = 5.7$ mm $\frac{c^*}{t_w} = \frac{225.2}{5.7} = 39.51 < 72\epsilon(= 72)$ **Class 1** Therefore, the section is **class 1**	Section class 1
6	6.2.6(3)	For I beam with load applied on flange, consider the case of rolled I sections with load parallel to web: Shear area, A_v $= A - 2bt_f + (t_w + 2r)t_f$ $= 28 \times 10^2 - 2(101.6)(6.8) + (5.7 + 2(7.6))(6.8)$ $= 1560.36$ mm^2	
	6.2.6(2)	$V_{pl,Rd} = \frac{A_v(f_y/\sqrt{3})}{\gamma_{M0}}$ $= \frac{1560.36 \times 235}{\sqrt{3}}$ $=$ **211.71 kN**	$V_{pl,Rd} = 211.71$ kN
7		$\frac{V_{Ed}}{V_{pl,Rd}} = \frac{28.88}{211.71} = \mathbf{0.14} < 1$ The shear resistance of the section is adequate	$\frac{V_{Ed}}{V_{pl,Rd}} = 0.14$
8	6.2.6(6)	Check for shear buckling failure: $h_w = d + 2r$ $= 225.2 + 2(7.6)$ $= 240.4$ mm $t_w = 5.7$ mm $\frac{h_w}{t_w} = \frac{240.4}{5.7} = 42.18 < 72\frac{\epsilon}{\eta}(= 72)$ Shear buckling check is not required	
9		*This step is skipped as shear buckling check is not required*	
10	6.2.5(2)	For Class 1 section, Bending moment resistance, $M_{c,Rd} = M_{pl,Rd}$ $= \frac{W_{pl,y} f_y}{\gamma_{M0}}$ $= \frac{259 \times 10^{-6} \times 235 \times 10^6}{1}$ $=$ **60.87 kNm**	$M_{c,Rd} = 60.87$ kNm

(continued)

(continued)

Step	Reference	Action/calculation	Conclusion
11		$\frac{M_{Ed}}{M_{c,Rd}} = \frac{36.08}{60.87} = \mathbf{0.59} < 1$ The bending resistance of the section is adequate	$\frac{M_{Ed}}{M_{c,Rd}} = 0.59$
12	6.2.8(2)	Check for combination of shear and bending failure: $\frac{V_{Ed}}{V_{pl,Rd}} = \frac{28.88}{211.71} = \mathbf{0.14} < 0.5$ Reduction in bending resistance is not required	
13		*This step is skipped as reduction in bending resistance is not required*	
14		*This step is skipped as reduction in bending resistance is not required*	
15		For SLS, partial factor of safety for both permanent action and variable action selected is 1.0 Serviceability load, w_{ser} $= 1.0G_k + 1.0Q_k$ $= 1.0(5.22) + 1.0(3)$ $= 8.22$ kN/m For simply supported beam, maximum deflection can be determined using equation below: Maximum deflection, Δ_{max} $= \frac{5wL^4}{384EI}$ $= \frac{5 \times 8.22 \times 10^3 \times 5^4}{384 \times 210 \times 10^9 \times 2841 \times 10^{-8}}$ $= 0.01121$ m $=$ **11.21 mm**	$\Delta_{max} = 11.21$ mm
16	NA2.23	Assume the beam carries plaster of other brittle finishes, Allowable deflection, Δ_{all} $= \frac{L}{360}$ $= \frac{5}{360}$ $= 0.01389$ m $=$ **13.89 mm**	$\Delta_{all} = 13.89$ mm
17		$\frac{\Delta_{max}}{\Delta_{all}} = \frac{11.21}{13.89} = \mathbf{0.81} < 1$ The deflection is allowable	$\frac{\Delta_{max}}{\Delta_{all}} = 0.81$
18		Check the following ratio: $\frac{V_{Ed}}{V_{pl,Rd}} = \frac{28.88}{211.71} = \mathbf{0.14}$ $\frac{M_{Ed}}{M_{c,Rd}} = \frac{36.08}{60.87} = \mathbf{0.59}$ $\frac{\Delta_{max}}{\Delta_{all}} = \frac{11.21}{13.89} = \mathbf{0.81}$ Although the value of $\frac{V_{Ed}}{V_{pl,Rd}}$ is significantly small, but the value of $\frac{M_{Ed}}{M_{c,Rd}}$ is greater than 0.5 and the value of $\frac{\Delta_{max}}{\Delta_{all}}$ is approaching 1. Therefore, the beam section 254 × 102 × 22 is **optimum**	

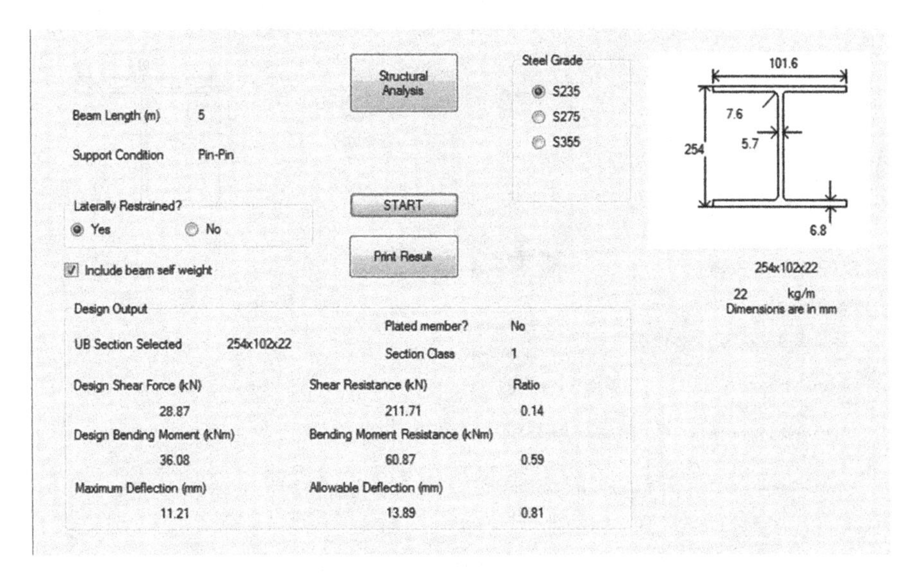

Fig. 2.4 Result for Example 2-1 using steel design based on EC3 program

2.2.3 *Example 2-2 Design of a Laterally Restrained Beam*

Check the suitability of a 305 × 102 × 25 section for a beam 7 m in length and subjected to a uniform load (Fig. 2.5). Use steel grade S235. Assume the beam is laterally restrained and sits on 100 mm bearings at each end. Take the self-weight of the beam into account (Fig. 2.6).

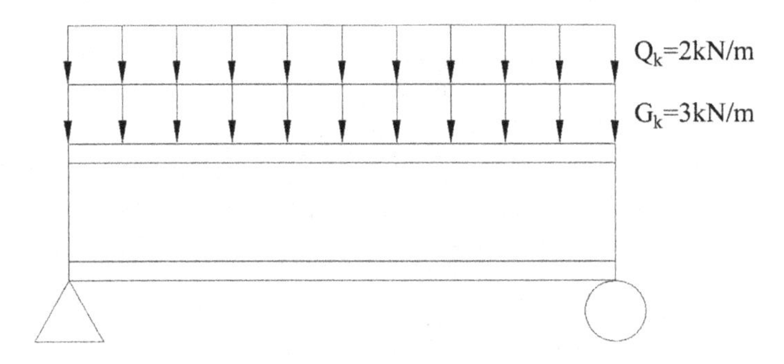

Fig. 2.5 Example 2-2

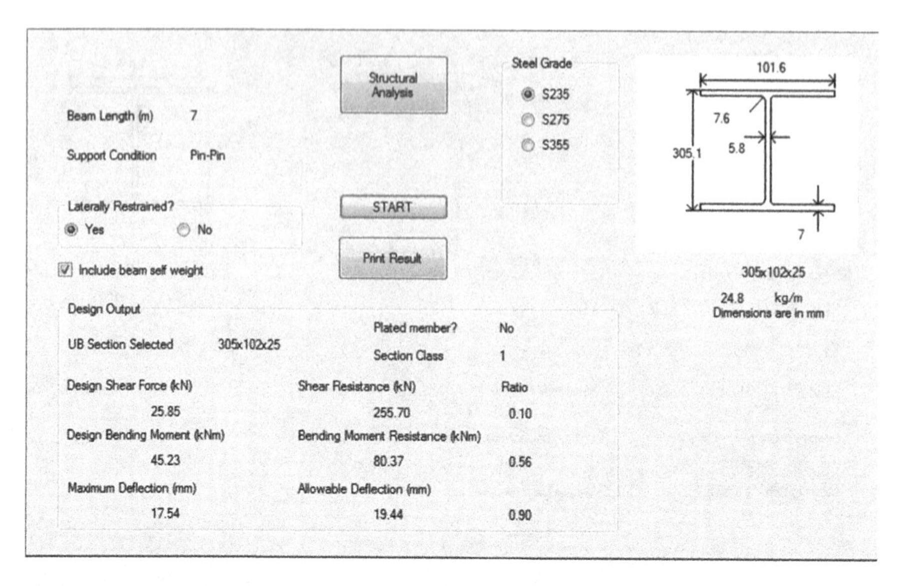

Fig. 2.6 Result for Example 2-2 using steel design based on EC3 program

Step	Reference	Action/calculation	Conclusion
1	References are to BS EN 1993-1-1 unless otherwise stated	From figure, the beam is **simply supported**	
2		Permanent action, $\boldsymbol{G_k}$ **= 3 kN/m** Variable action, $\boldsymbol{Q_k}$ **= 2 kN/m**	
3	Table 3.1	Steel grade = **S235** Assume the thicknesses of web and flange are less than 40 mm: $\boldsymbol{f_y}$ **= 235 N/mm²**	$f_y = 235$ N/mm^2
	BS 4 Part 1 2005	Try the following beam section: Select beam section **305 × 102 × 25** The properties of the section is as follows: Mass per meter = 24.8 kg/m Depth of section, $D = 305.1$ mm Width of section, $b = 101.6$ mm Thickness of web, $t_w = 5.8$ mm Thickness of flange, $t_f = 7.0$ mm Root radius, $r = 7.6$ mm Depth between fillets, $d = 275.9$ mm Second moment of area about major (*y-y*) axis, *Iy* = 4455 cm^4 Elastic modulus about major (*y-y*) axis, *Wel,y* = 292 cm^3 Plastic modulus about major (*y-y*) axis, *Wpl,y* = 342 cm^3 Area of section, $A = 31.6$ cm^2	

(continued)

(continued)

Step	Reference	Action/calculation	Conclusion
4		Self-weight of beam section = 24.8 kg/m × 9.81 N/kg = **0.24 kN/m** For ULS, partial factor of safety for both permanent action and variable action selected are 1.35 and 1.5 respectively Ultimate load, w_{ult} $= 1.35G_k + 1.5Q_k$ = 1.35(3 + 0.24) + 1.5(2) = **7.37 kN/m**	Design load = 7.37 kN/m
		For simply supported beam, V_{Ed} and M_{Ed} can be determined using equation below: V_{Ed} $= \frac{w_{ult}L}{2}$ $= \frac{7.37 \times 7}{2}$ = **25.82 kN**	V_{Ed} = 25.82 kN
		M_{Ed} $= \frac{w_{ult}L^2}{8}$ $= \frac{7.37 \times 7^2}{8}$ = **45.14 kNm**	M_{Ed} = 45.14 kNm
5	Table 5.2	Section classification: i. $f_y = 235$ N/mm^2 ε = 1 **Class 1** ii. Rolled section, outstand flange: $c = \frac{b - t_w - 2r}{2}$ $= \frac{101.6 - 5.8 - 2(7.6)}{2}$ = 40.30 mm $t_f = 7$ mm $\frac{c}{t_f} = \frac{40.30}{7} = 5.76 < 9\epsilon (= 9)$ **Class 1** iii. Rolled section, web with neutral axis at mid depth: $c^* = d$ = 275.9 mm $t_w = 5.8$ mm $\frac{c^*}{t_w} = \frac{275.9}{5.8} = 47.57 < 72\epsilon (= 72)$ **Class 1** Therefore, the section is **class 1**	Section class 1
6	6.2.6(3)	For I beam with load applied on flange, consider the case of rolled I sections with load parallel to web: Shear area, A_v $= A - 2bt_f + (t_w + 2r)t_f$ $= 31.6 \times 10^2 - 2(101.6)(7) + (5.8 + 2(7.6))(7)$ = 1884.60 mm^2	
	6.2.6(2)	$V_{pl,Rd} = \frac{A_v (f_y/\sqrt{3})}{\gamma_{M0}}$ $= \frac{1884.60 \times 235}{\sqrt{3}}$ = **255.70 kN**	$V_{pl,Rd}$ = 255.70 kN

(continued)

(continued)

Step	Reference	Action/calculation	Conclusion
7		$\frac{V_{Ed}}{V_{pl,Rd}} = \frac{25.82}{255.70} = \mathbf{0.10} < 1$ The shear resistance is adequate	$\frac{V_{Ed}}{V_{pl,Rd}} = 0.10$
8	6.2.6(6)	Check for shear buckling failure: $h_w = d + 2r$ $= 275.9 + 2(7.6)$ $= 291.1$ mm $t_w = 5.8$ mm $\frac{h_w}{t_w} = \frac{291.1}{5.8} = 50.19 < 72\frac{\epsilon}{\eta}(= 72)$ Shear buckling check is not required	
9		*This step is skipped as shear buckling check is not required*	
10	6.2.5(2)	For Class 1 section, Bending moment resistance, $M_{c,Rd} = M_{pl,Rd}$ $= \frac{W_{pl}f_y}{\gamma_{M0}}$ $= \frac{342 \times 10^{-6} \times 235 \times 10^6}{1}$ $= \mathbf{80.37\ kNm}$	$M_{c,Rd} = 80.37$ kNm
11		$\frac{M_{Ed}}{M_{c,Rd}} = \frac{45.14}{80.37} = \mathbf{0.56} < 1$ The bending resistance of the section is adequate	$\frac{M_{Ed}}{M_{c,Rd}} = 0.56$
12	6.2.8(2)	Check for combination of shear and bending failure: $\frac{V_{Ed}}{V_{pl,Rd}} = \frac{25.82}{255.70} = 0.10 < 0.5$ Reduction in bending resistance is not required	
13		*This step is skipped as reduction in bending resistance is not required*	
14		*This step is skipped as reduction in bending resistance is not required*	
15		For SLS, partial factor of safety or both permanent action and variable action selected is 1.0. Serviceability load, w_{ser} $= 1.0G_k + 1.0Q_k$ $= 1.0(3.24) + 1.0(2)$ $= 5.24$ kN/m For simply supported beam, maximum deflection can be determined using equation below: Maximum deflection, Δ_{max} $= \frac{5wL^4}{384EI}$ $= \frac{5 \times 5.24 \times 10^3 \times 7^4}{384 \times 210 \times 10^9 \times 4455 \times 10^{-8}}$ $= 0.01751$ m $= \mathbf{17.51\ mm}$	$\Delta_{max} = 17.51$ mm

(continued)

(continued)

Step	Reference	Action/calculation	Conclusion
16	NA2.23	Assume the beam carries plaster of other brittle finishes, Allowable deflection, Δ_{all} $= \frac{L}{360}$ $= \frac{7}{360}$ $= 0.01944$ m $= \mathbf{19.44\ mm}$	Δ_{all} = 19.44 mm
17		$\frac{\Delta_{max}}{\Delta_{all}} = \frac{17.51}{19.44} = \mathbf{0.90} < 1$ The deflection is allowable	$\frac{\Delta_{max}}{\Delta_{all}} = 0.90$
18		Check the following ratio: $\frac{V_{Ed}}{V_{pl,Rd}} = \frac{25.82}{255.70} = \mathbf{0.10}$ $\frac{M_{Ed}}{M_{c,Rd}} = \frac{45.14}{80.37} = \mathbf{0.56}$ $\frac{\Delta_{max}}{\Delta_{all}} = \frac{17.51}{19.44} = \mathbf{0.90}$ The section is suitable for the condition. Other than that, the value of $\frac{\Delta_{max}}{\Delta_{all}}$ is approaching 1, while the value of $\frac{M_{Ed}}{M_{c,Rd}}$ is 0.5. Therefore, the beam section 305 × 102 × 25 is **optimum**	

2.2.4 Example 2-3 Design of a Laterally Restrained Beam

Check the suitability of a 305 × 102 × 28 section for the propped cantilever beam 8 m in length and subjected to a uniform load (Fig. 2.7). Use steel grade S235, and assume the beam is laterally restrained. Ignore the self-weight of the beam. If the said section is not suitable, briefly describe the action to be taken to make the section suitable for this condition.

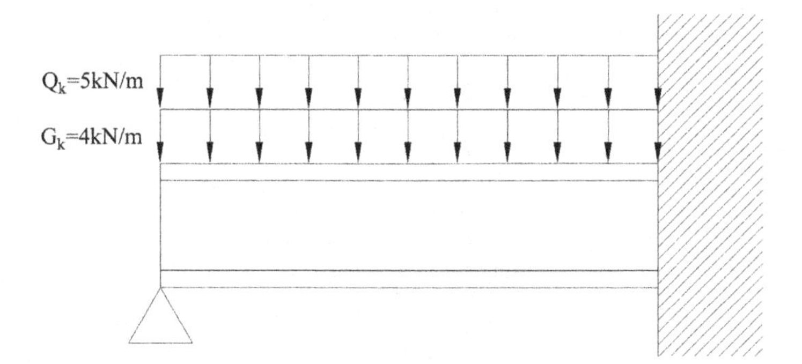

Fig. 2.7 Example 2-3

Step	Reference	Action/calculation	Conclusion
1	References are to BS EN 1993-1-1 unless otherwise stated	From figure, the support condition of beam is **fixed-pinned**	
2		Permanent action, G_k = **4 kN/m** Variable action, Q_k = **5 kN/m**	
3	Table 3.1	Steel grade = **S235** Assume the thicknesses of web and flange are less than 40 mm: f_y = **235 N/mm^2**	f_y = 235 N/mm^2
	BS 4 Part 1 2005	Try the following beam section: Select beam section **305 × 102 × 28** The properties of the section is as follows: Mass per meter = 28.2 kg/m Depth of section, D = 308.7 mm Width of section, b = 101.8 mm Thickness of web, t_w = 6.0 mm Thickness of flange, t_f = 8.8 mm Root radius, r = 7.6 mm Depth between fillets, d = 275.9 mm Second moment of area about major (*y-y*) axis, *Iy* = 5366 cm^4 Elastic modulus about major (*y-y*) axis, *Wel,y* = 348 cm^3 Plastic modulus about major (*y-y*) axis, *Wpl,y* = 403 cm^3 Area of section, A = 35.9 cm^2	
4		For ULS, partial factor of safety for both permanent action and variable action selected are 1.35 and 1.5 respectively Ultimate load, w_{ult} $= 1.35G_k + 1.5Q_k$ $= 1.35(4) + 1.5(5)$ = **12.90 kN/m**	Design load = 12.90 kN/m
		For propped cantilever (beam with fixed-pinned support condition), V_{Ed} and M_{Ed} can be determined using equation below: V_{Ed} $= \frac{5w_{ult}L}{8}$ $= \frac{5\times 12.90\times 8}{8}$ = **64.50 kN**	V_{Ed} = 64.50 kN

(continued)

(continued)

Step	Reference	Action/calculation	Conclusion
		M_{Ed} $= \frac{w_{ult}L^2}{8}$ $= \frac{12.90 \times 8^2}{8}$ $=$ **103.20 kNm**	$M_{Ed} = 103.20$ kNm
5	Table 5.2	Section classification: i. $f_y = 235$ N/mm^2 $\varepsilon = 1$ **Class 1** ii. Rolled section, outstand flange: $c = \frac{b - t_w - 2r}{2}$ $= \frac{101.8 - 6 - 2(7.6)}{2}$ $= 40.3$ mm $t_f = 8.8$ mm $\frac{c}{t_f} = \frac{40.3}{8.8} = 4.58 < 9\epsilon (= 9)$ **Class 1** iii. Rolled section, web with neutral axis at mid depth: $c^* = d$ $= 275.9$ mm $t_w = 6$ mm $\frac{c^*}{t_w} = \frac{275.9}{6} = 45.98 < 72\epsilon (= 72)$ **Class 1** Therefore, the section is **class 1**	Section class 1
6	6.2.6(3)	For I beam with load applied on flange, consider the case of rolled I sections with load parallel to web: Shear area, A_v $= A - 2bt_f + (t_w + 2r)t_f$ $= 35.9 \times 10^2 - 2(101.8)$ $(8.8) + (6 + 2(7.6))(8.8)$ $= 1984.88$ mm^2	
	6.2.6(2)	$V_{pl,Rd} = \frac{A_v(f_y/\sqrt{3})}{\gamma_{M0}}$ $= \frac{1984.88 \times 235}{\sqrt{3}}$ $=$ **269.30 kN**	$V_{pl,Rd} = 269.30$ kN
7		$\frac{V_{Ed}}{V_{pl,Rd}} = \frac{64.50}{269.30} = \mathbf{0.24} < 1$ The shear resistance is adequate	$\frac{V_{Ed}}{V_{pl,Rd}} = 0.24$
8	6.2.6(6)	Check for shear buckling failure: $h_w = d + 2r$ $= 275.9 + 2(7.6)$ $= 291.1$ mm $t_w = 6$ mm $\frac{h_w}{t_w} = \frac{291.1}{6} = 48.52 < 72\frac{\epsilon}{\eta} (= 72)$ Shear buckling check is not required	

(continued)

(continued)

Step	Reference	Action/calculation	Conclusion
9		*This step is skipped as shear buckling check is not required*	
10	6.2.5(2)	For Class 1 section, Bending moment resistance, $M_{c,Rd} = M_{pl,Rd}$ $= \frac{W_{pl}f_y}{\gamma_{M0}}$ $= \frac{403 \times 10^{-6} \times 235 \times 10^6}{1}$ $=$ **94.71 kNm**	$M_{c,Rd} = 94.71$ kNm
11		$\frac{M_{Ed}}{M_{c,Rd}} = \frac{103.20}{94.71} = \mathbf{1.09} > 1$ The bending resistance of the section is not adequate	$\frac{M_{Ed}}{M_{c,Rd}} = 1.09$

The section specified is **not suitable** for the situation. Besides selecting a larger section, higher-grade steel such as grade S275 can be used.

Step	Reference	Action/calculation	Conclusion
3		Steel grade = **S275** The thicknesses of web and flange are 6.0 mm and 8.8 mm, which are less than 40 mm f_y = **275 N/mm^2**	f_y = 275 N/mm^2
	BS 4 Part 1 2005	Use beam section **305** × **102** × **28** The properties of the section is as follows: Mass per meter = 28.2 kg/m Depth of section, D = 308.7 mm Width of section, b = 101.8 mm Thickness of web, t_w = 6.0 mm Thickness of flange, t_f = 8.8 mm Root radius, r = 7.6 mm Depth between fillets, d = 275.9 mm Second moment of area about major (y-y) axis, Iy = 5366 cm^4 Elastic modulus about major (y-y) axis, Wel,y = 348 cm^3 Plastic modulus about major (y-y) axis, Wpl,y = 403 cm^3 Area of section, A = 35.9 cm^2	
4		From previous calculation, Ultimate load, w_{ult} = **12.90 kN/m**	Design load = 12.90 kN/m
		V_{Ed} = **64.50 kN**	V_{Ed} = 64.50 kN
		M_{Ed} = **103.20 kNm**	M_{Ed} = 103.20 kNm

(continued)

(continued)

Step	Reference	Action/calculation	Conclusion
5	Table 3.1	Section classification: i. $f_y = 275$ N/mm^2 $\varepsilon = 0.92$ **Class 2** ii. Rolled section, outstand flange: $c = \frac{b-t_w-2r}{2}$ $= \frac{101.8-6-2(7.6)}{2}$ $= 40.3$ mm $t_f = 8.8$ mm $\frac{c}{t_f} = \frac{40.3}{8.8} = 4.58 < 9\epsilon(= 8.28)$ **Class 1** iii. Rolled section, web with neutral axis at mid depth: $c^* = d$ $= 275.9$ mm $t_w = 6$ mm $\frac{c^*}{t_w} = \frac{275.9}{6} = 45.98 < 72\epsilon(= 66.24)$ **Class 1** Therefore, the section is **class 2**	Section class 2
6	6.2.6(3)	For I beam with load applied on flange, consider the case of rolled I sections with load parallel to web: Shear area, A_v $= A - 2bt_f + (t_w + 2r)t_f$ $= 35.9 \times 10^2 - 2(101.8)(8.8) + (6 + 2(7.6))$ (8.8) $= 1984.88$ mm^2	
	6.2.6(2)	$V_{pl,Rd} = \frac{A_v\left(f_y/\sqrt{3}\right)}{\gamma_{M0}}$ $= \frac{1984.88 \times 275}{\sqrt{3}}$ $=$ **315.14 kN**	$V_{pl,Rd} = 315.14$ kN
7		$\frac{V_{Ed}}{V_{pl,Rd}} = \frac{64.50}{315.14} = \mathbf{0.20} < 1$ The shear resistance is adequate	$\frac{V_{Ed}}{V_{pl,Rd}} = 0.20$
8	6.2.6(6)	Check for shear buckling failure: $h_w = d + 2r$ $= 275.9 + 2(7.6)$ $= 291.1$ mm $t_w = 6$ mm $\frac{h_w}{t_w} = \frac{291.1}{6} = 48.52 < 72\frac{\epsilon}{\eta}(= 72)$ Shear buckling check is not required	

(continued)

(continued)

Step	Reference	Action/calculation	Conclusion
9		*This step is skipped as shear buckling check is not required*	
10	6.2.5(2)	For Class 2 section, Bending moment resistance, $M_{c,Rd} = M_{pl,Rd}$ $= \frac{W_{pl}f_y}{\gamma_{M0}}$ $= \frac{403 \times 10^{-6} \times 275 \times 10^6}{1}$ $=$ **110.83 kNm**	$M_{c,Rd} = 110.83$ kNm
11		$\frac{M_{Ed}}{M_{c,Rd}} = \frac{103.20}{110.83} = \mathbf{0.93} < 1$ The bending resistance of the section is adequate	$\frac{M_{Ed}}{M_{c,Rd}} = 0.93$
12	6.2.8(2)	Check for combination of shear and bending failure: $\frac{V_{Ed}}{V_{pl,Rd}} = \frac{64.50}{315.14} = \mathbf{0.20} < 0.5$ Reduction in bending resistance is not required	
13		*This step is skipped as reduction in bending resistance is not required*	
14		*This step is skipped as reduction in bending resistance is not required*	
15		For SLS, partial factor of safety or both permanent action and variable action selected is 1.0. Serviceability load, w_{ser} $= 1.0G_k + 1.0Q_k$ $= 1.0(4) + 1.0(5)$ $= 9$ kN/m For propped cantilever, maximum deflection can be determined using equation below: Maximum deflection, Δ_{max} $= \frac{wL^4}{185EI}$ $= \frac{9 \times 10^3 \times 8^4}{185 \times 210 \times 10^9 \times 5366 \times 10^{-8}}$ $= 0.01768$ m $=$ **17.68 mm**	$\Delta_{max} = 17.68$ mm
16	NA2.23	Assume the beam carries plaster of other brittle finishes, Allowable deflection, Δ_{all} $= \frac{L}{360}$ $= \frac{8}{360}$ $= 0.02222$ m $=$ **22.22 mm**	$\Delta_{all} = 22.22$ mm

(continued)

(continued)

Step	Reference	Action/calculation	Conclusion
17		$\frac{\Delta_{max}}{\Delta_{all}} = \frac{17.68}{22.22} = 0.80 < 1$ The deflection is allowable	$\frac{\Delta_{max}}{\Delta_{all}} = 0.80$
18		Check the following ratio: $\frac{V_{Ed}}{V_{pl,Rd}} = \frac{64.50}{315.14} = \mathbf{0.20}$ $\frac{M_{Ed}}{M_{c,Rd}} = \frac{103.20}{110.83} = \mathbf{0.93}$ $\frac{\Delta_{max}}{\Delta_{all}} = \frac{17.68}{22.22} = \mathbf{0.80}$ By increase the steel grade, the beam section become adequate. The values of $\frac{M_{Ed}}{M_{c,Rd}}$ and $\frac{\Delta_{max}}{\Delta_{all}}$ are approaching 1. Therefore, the beam section 305 × 102 × 28 is **optimum**	

2.3 Design Procedure for a Laterally Unrestrained Beam

The design procedure for a laterally unrestrained beam is as follows:

1. Determine the support condition (i.e., pin, roller, or fixed at both ends of the beam).
2. Determine the DL and LL that act on the beam.
3. Choose the steel grade (refer to Table 2.1). Refer to BS 4 Part 1 2005 to choose the beam section for use in construction. A table for the universal beam section and its corresponding properties is provided in Appendix A.2.
4. Perform a structural analysis to determine the maximum shear force V_{Ed} and bending moment M_{Ed} induced by loading. Prior to the analysis, the partial safety factor for ULS (Table 1.1) is applied to the actions determined in Step 2, including the self-weight of the beam section.
5. Classify the beam section (refer to Table 2.2).
6. Determine the critical buckling moment using the equation below. The support condition influences the effective length of the member subjected to buckling, as shown in Table 2.5 (Refer to Appendix A.2 for the section properties of the beam sections).

$$M_{cr} = \frac{\pi^2 EI_z}{(KL)^2}\sqrt{\left(\frac{I_w}{I_z} + \frac{(KL)^2 GI_t}{\pi^2 EI_z}\right)} \tag{2.5}$$

Table 2.5 Values of effective length factor K for different support conditions (BS5950: Part 1 4.7.10)

Support condition	Effective length factor, K
Fixed-fixed	0.7
Fixed-pinned	0.85
Pinned-pinned	1.0
Fixed-free	2.0

where

E is modulus of elasticity of steel = 210×10^9 N/m^2
I_z is second moment of area about z-z axis by referring to Appendix A.2
K is effective length factor obtained from Step 6 (Table 2.5)
L is length of beam
I_w is warping constant by referring to Appendix A.2
G is shear modulus of steel = 81×10^9 N/m^2
I_t is torsional constant by referring to Appendix A.2

(SN003b Access Steel document)

7. Determine the slenderness for lateral torsional buckling $\bar{\lambda}_{LT}$ using the equation below.

$$\bar{\lambda}_{LT} = \begin{cases} \sqrt{\frac{W_{pl,y} f_y}{M_{cr}}}, \textit{Class 1 and 2 sections} \\ \sqrt{\frac{W_{el,y} f_y}{M_{cr}}}, \textit{Class 3 sections} \\ \sqrt{\frac{W_{eff,y} f_y}{M_{cr}}}, \textit{Class 4 sections} \end{cases} \tag{2.6}$$

where

$W_{pl,y}$ is plastic section modulus about y-y axis by referring to Appendix A.2
$W_{el,y}$ is elastic section modulus about y-y axis by referring to Appendix A.2
$W_{eff,y}$ is effective section modulus about y-y axis
f_y is yield strength of steel obtained from Step 3 (Table 2.1)
M_{cr} is critical buckling moment obtained from Step 6 (Eq. 2.5)

(BS EN 1993-1-1:2005 6.3.2.2(1))

8. Determine the imperfection factors for lateral-torsional buckling, α_{LT} and ϕ_{LT}. These values may be determined using two approaches: general case approach, which is applicable to all section types, and rolled section approach, which is

Table 2.6 Values of the imperfection factor α_{LT} for different approaches (BS EN 1993-1-1:2005 Tables 6.3, 5.2, and 5.2)

Rolled I section			
"General case" approach		"Rolled section" approach	
Limit	α_{LT}	Limit	α_{LT}
$h/b \leq 2$	0.21	$h/b \leq 2$	0.34
$h/b > 2$	0.34	$2 < h/b \leq 3.1$	0.49
		$h/b > 3.1$	0.76

Where h is depth of section by referring to Appendix A.2
b is width of section by referring to Appendix A.2

only applicable to rolled sections. The depth of the section is denoted by h. Both approaches may generate values with significant differences.

$$\phi_{LT} = \begin{cases} 0.5\left[1 + \alpha_{LT}\left(\bar{\lambda}_{LT} - 0.2\right) + \bar{\lambda}_{LT}^2\right], \text{“\textit{General Case}” \textit{approach}} \\ 0.5\left[1 + \alpha_{LT}\left(\bar{\lambda}_{LT} - 0.4\right) + 0.75\bar{\lambda}_{LT}^2\right], \text{“\textit{Rolled Section}” \textit{approach}} \end{cases} \tag{2.7}$$

where

α_{LT} is imperfection factor obtained from Step 8 (Table 2.6)
$\bar{\lambda}_{LT}$ is slenderness for lateral torsional buckling obtained from Step 7 (Eq. 2.6)

(BS EN 1993-1-1:2005 6.3.2.2(1) and 6.3.2.3(1))

9. Determine the lateral torsional buckling reduction factor χ_{LT}. In case the rolled section approach is used, refer to Table 2.7.

$$\chi_{LT} = \frac{1}{\phi_{LT} + \sqrt{\phi_{LT}^2 - \bar{\lambda}_{LT}^2}}, \text{“\textit{General Case}” \textit{approach}}$$

For "Rolled Section" approach

$$\begin{aligned} \chi_{LT} &= \frac{1}{\phi_{LT} + \sqrt{\phi_{LT}^2 - 0.75\bar{\lambda}_{LT}^2}}, \chi_{LT} \leq 1 \textit{ and } \chi_{LT} \leq \frac{1}{\bar{\lambda}_{LT}^2} \\ f &= 1 - 0.5(1 - K_c)\left[1 - 2\left(\bar{\lambda}_{LT} - 0.8\right)^2\right] \leq 1 \\ \chi_{LT,mod} &= \frac{\chi_{LT}}{f} \leq 1 \end{aligned} \tag{2.8}$$

where

ϕ_{LT} is obtained from Step 8 (Eq. 2.7)
$\bar{\lambda}_{LT}$ is slenderness for lateral torsional buckling obtained from Step 7 (Eq. 2.6)
K_C is correlation factor for moment distribution obtained from Step 9 (Table 2.7)

(BS EN 1993-1-1:2005 6.3.2.2(1) and 6.3.2.3(1))

10. Determine the buckling moment resistance. When the rolled section approach is used in the previous steps, $\chi_{LT,mod}$ should be used instead of χ_{LT} in the following equation. γ_{M1} should be set as 1.0.

$$M_{b,Rd} = \begin{cases} \chi_{LT} W_{pl,y} \frac{f_y}{\gamma_{M1}}, \textit{Class 1 and 2 sections} \\ \chi_{LT} W_{el,y} \frac{f_y}{\gamma_{M1}}, \textit{Class 3 sections} \\ \chi_{LT} W_{eff,y} \frac{f_y}{\gamma_{M1}}, \textit{Class 4 sections} \end{cases} \tag{2.9}$$

Table 2.7 Correlation between moment distribution and K_c (BS EN 1993-1-1:2005 Table 6.6)

Moment distribution	K_C
 $\psi = 1$	1
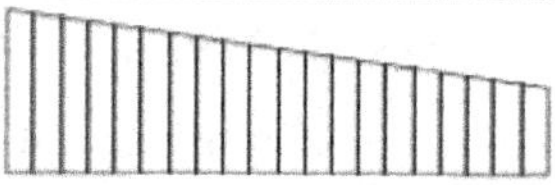 $-1 \leq \psi \leq 1$	$\frac{1}{1.33-0.33\psi}$
	0.94
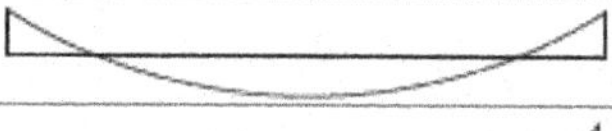	0.90
	0.91
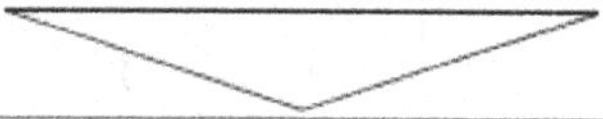	0.86
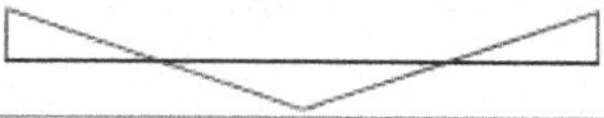	0.77
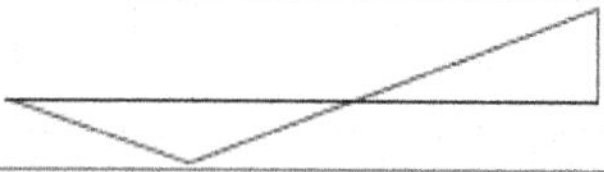	0.82

Where ψ is the ratio of moment at two ends

where

$W_{pl,y}$ is plastic section modulus about y-y axis by referring to Appendix A.2
$W_{el,y}$ is elastic section modulus about y-y axis by referring to Appendix A.2
$W_{eff,y}$ is effective section modulus about y-y axis
f_y is yield strength of steel obtained from Step 3 (Table 2.1)
χ_{LT} is lateral torsional buckling reduction factor obtained from Step 9 (Eq. 2.8)

(BS EN 1993-1-1:2005 6.3.2.1(3))

11. Compare the design bending moment of the structure and the buckling moment resistance of the section. If the buckling moment resistance of the structure is insufficient, repeat Step 3 to choose a better section. Otherwise, proceed to Step 12.
12. Determine the shear resistance of the section by referring to Table 2.3 and Eq. 2.1.
13. Compare the design shear force on the structure and the shear resistance of the section. If the shear resistance of the structure is insufficient, repeat Step 3 to choose a better section. Otherwise, proceed to Step 14.
14. Determine the maximum deflection of the structure under the loading specified in Step 2. The load combination for this calculation should be any of those specified for the SLS design, as shown in Table 1.1.
15. Determine the allowable deflection of the structure by referring to Table 2.4.
16. Compare the maximum deflection and allowable deflection of the structure. If the deflection of the structure exceeds the allowable deflection, repeat Step 3 to choose a better section. Otherwise, proceed to Step 17.
17. Check whether the section is an overdesign by checking the ratio of design value to resistance for shear and bending and the ratio of maximum deflection to allowable deflection. If both ratios are less than 0.5, repeat Step 3 and choose a smaller section to ensure optimum design.

2.3.1 Design Flowchart for a Laterally Unrestrained Beam

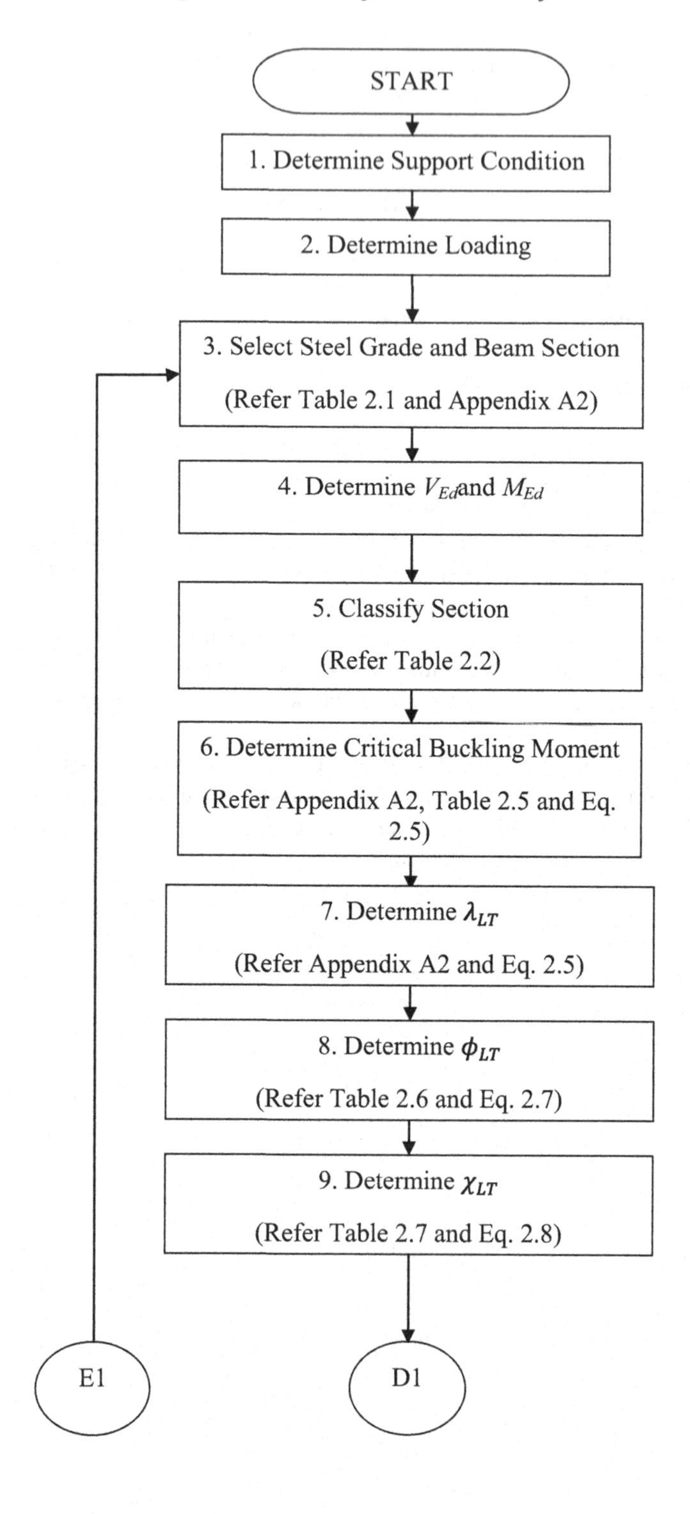

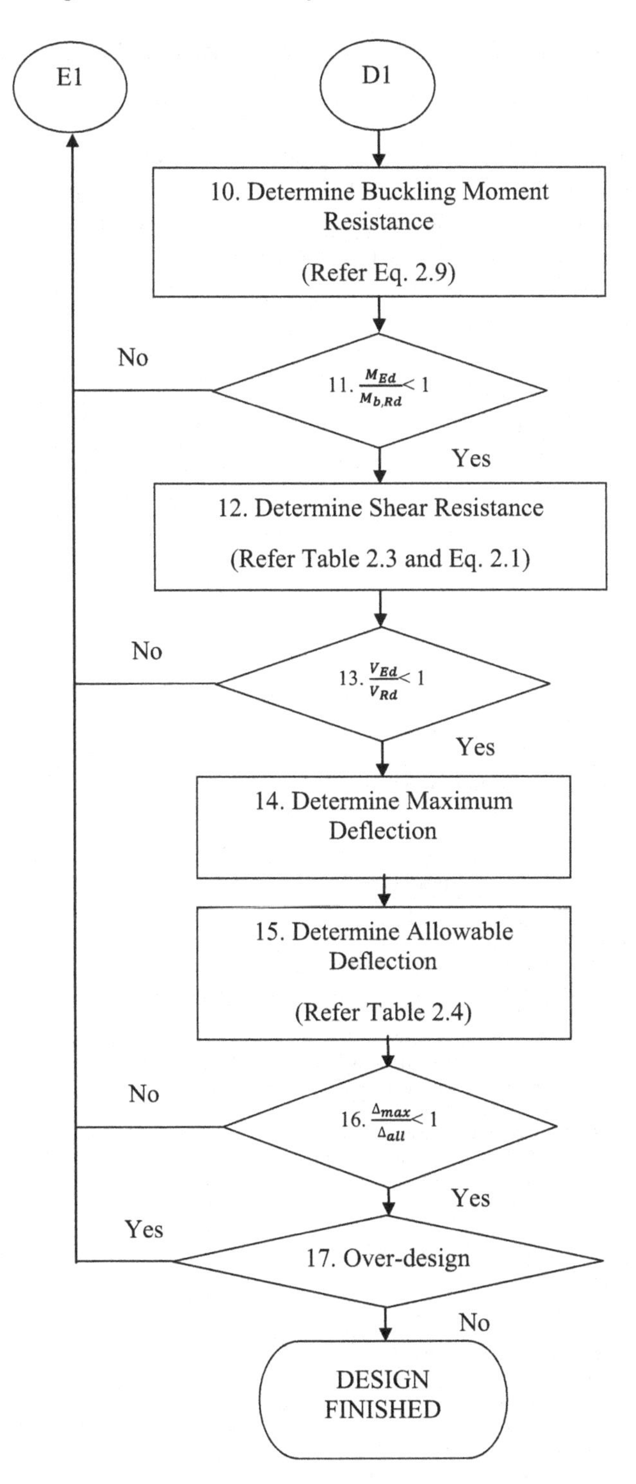
E1
D1
10. Determine Buckling Moment Resistance
(Refer Eq. 2.9)
No
11. $\frac{M_{Ed}}{M_{b,Rd}} < 1$
Yes
12. Determine Shear Resistance
(Refer Table 2.3 and Eq. 2.1)
No
13. $\frac{V_{Ed}}{V_{Rd}} < 1$
Yes
14. Determine Maximum Deflection
15. Determine Allowable Deflection
(Refer Table 2.4)
No
16. $\frac{\Delta_{max}}{\Delta_{all}} < 1$
Yes
Yes
17. Over-design
No
DESIGN FINISHED

2.3.2 Example 2-4 Design of a Laterally Unrestrained Beam

Check the suitability of a 457 × 191 × 89 section for a beam 10 m in length and subjected to a uniform load (Fig. 2.8). Use steel grade S235. Assume the beam is laterally unrestrained and sits on 100 mm bearings at each end. Ignore the self-weight of the beam. If the said section is not suitable, briefly describe the action to be taken to make the section suitable for this condition.

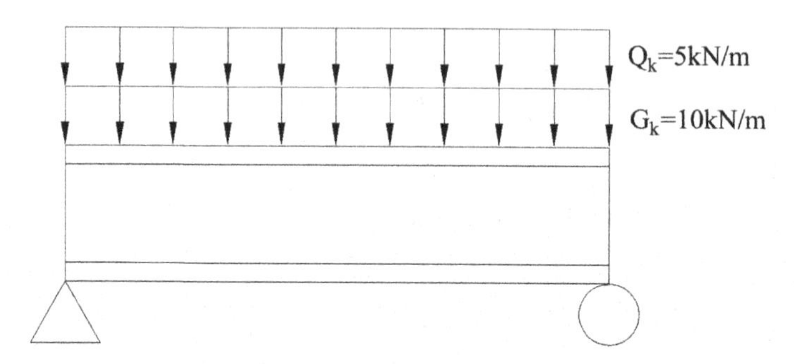

Fig. 2.8 Example 2-4

Step	Reference	Action/calculation	Conclusion
1	References are to BS EN 1993-1-1 unless otherwise stated	From figure, the beam is **simply supported**	
2		Permanent action, G_k = **10 kN/m** Variable action, Q_k = **5 kN/m**	
3	Table 3.1	Steel grade = **S235** Assume the thicknesses of web and flange are less than 40 mm: f_y = **235 N/mm²**	$f_y = 235\ \text{N/mm}^2$
	BS 4 Part 1 2005	Try the following beam section: Select beam section **457 × 191 × 89** The properties of the section is as follows: Mass per meter = 89.3 kg/m Depth of section, D = 463.4 mm Width of section, b = 191.9 mm Thickness of web, t_w = 10.5 mm Thickness of flange, t_f = 17.7 mm Root radius, r = 10.2 mm Depth between fillets, d = 407.6 mm Second moment of area about major (y-y) axis, Iy = 41020 cm⁴ Second moment of area about minor (z-z) axis, Iz = 2089 cm⁴	

(continued)

(continued)

Step	Reference	Action/calculation	Conclusion
		Elastic modulus about major (*y-y*) axis, *Wel,y* = 1770 cm^3 Plastic modulus about major (*y-y*) axis, *Wpl,y* = 2014 cm^3 Warping constant, I_w = 1.04 dm^6 Torsional constant, I_t = 90.7 cm^4 Area of section, A = 114 cm^2	
4		For ULS, partial factor of safety for both permanent action and variable action selected are 1.35 and 1.5 respectively Ultimate load, w_{ult} $= 1.35G_k + 1.5Q_k$ $= 1.35(10) + 1.5(5)$ **= 21.00 kN/m**	Design load = 21.00 kN/m
		For simply supported beam, V_{Ed} and M_{Ed} can be determined using equation below: V_{Ed} $= \frac{w_{ult}L}{2}$ $= \frac{21\times10}{2}$ **= 105.00 kN**	V_{Ed} = 105.00 kN
		M_{Ed} $= \frac{w_{ult}L^2}{8}$ $= \frac{21\times10^2}{8}$ **= 262.50 kNm**	M_{Ed} = 262.50 kNm
5	Table 5.2	Section classification: i. f_y = 235 N/mm^2 ε = 1 **Class 1** ii. Rolled section, outstand flange: $c = \frac{b-t_w-2r}{2}$ $= \frac{191.9 - 10.5 - 2(10.2)}{2}$ = 80.50 mm t_f = 17.7 mm $\frac{c}{t_f} = \frac{80.50}{17.7} = 4.55 < 9\epsilon(= 9)$ **Class 1** iii. Rolled section, web with neutral axis at mid depth: c^* = d = 407.6 mm t_w = 10.5 mm $\frac{c^*}{t_w} = \frac{407.6}{10.5} = 38.82 < 72\epsilon(= 72)$ **Class 1** Therefore, the section is **class 1**	Section class 1
6	SN003b access steel document	Critical buckling resistance can be determined using equation below. For simply supported beam, effective length factor, *K* is taken as 1.0: $M_{cr} = \frac{\pi^2 EI_z}{(KL)^2}\sqrt{\left(\frac{I_w}{I_z} + \frac{(KL)^2 GI_t}{\pi^2 EI_z}\right)}$ $= \frac{\pi^2 \times 210 \times 10^9 \times 2089 \times 10^{-8}}{(1.0 \times 10)^2}$ $\times\sqrt{\left(\frac{1.04 \times 10^{-6}}{2089 \times 10^{-8}} + \frac{(1.0 \times 10)^2 \times 81 \times 10^9 \times 90.7 \times 10^{-8}}{\pi^2 \times 210 \times 10^9 \times 2089 \times 10^{-8}}\right)}$	M_{cr} = 202.83 kNm

(continued)

(continued)

Step	Reference	Action/calculation	Conclusion
		$= \mathbf{202.83\ kNm}$	
7	6.3.2.2(1)	For Class 1 section, slenderness for lateral torsional buckling can be determined using equation below: $\bar{\lambda}_{LT} = \sqrt{\frac{W_{pl,y} f_y}{M_{cr}}}$ $= \sqrt{\frac{2014 \times 10^{-6} \times 235 \times 10^{6}}{202.83 \times 10^{3}}}$ $= \mathbf{1.53}$	$\bar{\lambda}_{LT} = 1.53$
8	Table 6.3 Table 6.4	$\frac{h}{b} = \frac{D}{b} = \frac{463.4}{191.9} = 2.41$ Determine imperfection factor using "General Case" approach: $\frac{h}{b} = 2.41 > 2$ $\alpha_{LT} = 0.34$ $\phi_{LT} = 0.5\left[1 + \alpha_{LT}(\bar{\lambda}_{LT} - 0.2) + \bar{\lambda}_{LT}^2\right]$ $= 0.5\left[1 + 0.34 \times (1.53 - 0.2) + (1.53)^2\right]$ $= \mathbf{1.89}$	$\phi_{LT} = 1.89$
9	6.3.2.2(1)	Lateral torsional buckling reduction factor can be determined using equation below: $\chi_{LT} = \frac{1}{\phi_{LT} + \sqrt{\phi_{LT}^2 - \bar{\lambda}_{LT}^2}}$ $= \frac{1}{1.89 + \sqrt{(1.89)^2 - (1.53)^2}}$ $= \mathbf{0.33}$	$\chi_{LT} = 0.33$
10	6.3.2.1(3)	For Class 1 section, $M_{b,Rd} = \chi_{LT} W_{pl,y} \frac{f_y}{\gamma_{M1}}$ $= \frac{0.33 \times 2014 \times 10^{-6} \times 235 \times 10^{6}}{1.0}$ $= \mathbf{156.18\ kNm}$	$M_{b,Rd} = 156.18$ kNm
11		$\frac{M_{Ed}}{M_{b,Rd}} = \frac{262.50}{156.18} = \mathbf{1.68} > 1$ The bending resistance of the section is not adequate	$\frac{M_{Ed}}{M_{b,Rd}} = 1.68$

The section specified is **not suitable** for the situation. Besides selecting a larger section, higher-grade steel may be selected **or** the buckling length of the beam may be reduced by providing a secondary beam or support at the mid-span of the beam (Fig. 2.9).

From the program, the optimum section for beam subjected to condition as specified in Example 2-4 is 533 × 210 × 122. This section is obviously larger than proposed 457 × 191 × 89 section. Therefore, the proposed section is inadequate.

2.3.3 Example 2-5 Design of a Laterally Unrestrained Beam

A secondary beam is connected to the mid-span of the primary beam by shear connection. The reaction force of the secondary beam is 30 kN. Select the optimum section for the primary beam 10 m in length (Fig. 2.10). Use steel grade S235.

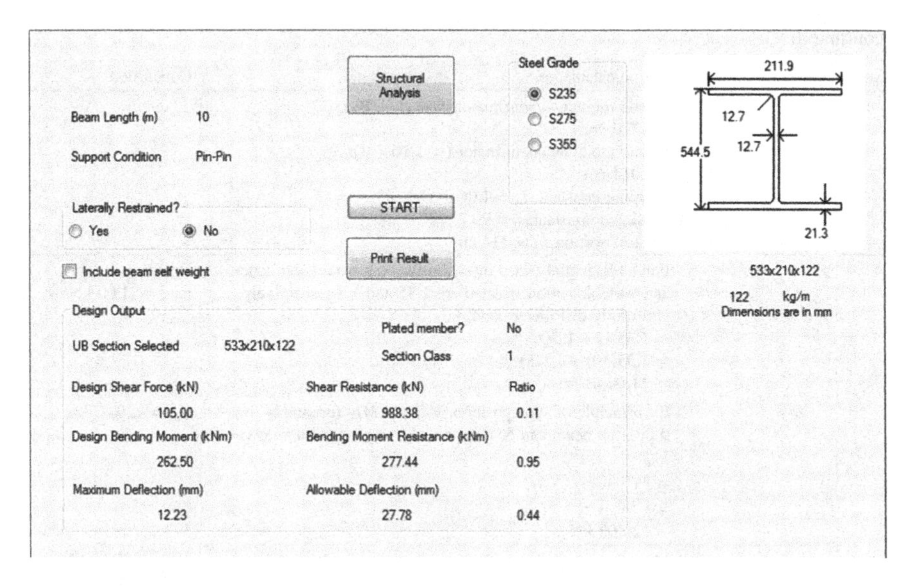

Fig. 2.9 Result for Example 2-4 using steel design based on EC3 program

Assume the primary beam is laterally unrestrained and sits on 100 mm bearings at each end. Ignore the self-weight of the beam.

Step	Reference	Action/calculation	Conclusion
1	References are to BS EN 1993-1-1 unless otherwise stated	From figure, the beam is **simply supported**	
2		Permanent action, $\boldsymbol{G_k}$ = **10 kN/m** Variable action, $\boldsymbol{Q_k}$ = **5 kN/m**	
3	Table 3.1	Steel grade = **S235** Assume the thicknesses of web and flange are less than 40 mm: $\boldsymbol{f_y = 235\ N/mm^2}$	$f_y = 235\ N/mm^2$
	BS 4 Part 1 2005	Randomly choose a beam section for the first trial: Select beam section **457** × **191** × **89** The properties of the section is as follows: Mass per meter = 89.3 kg/m Depth of section, D = 463.4 mm Width of section, b = 191.9 mm Thickness of web, t_w = 10.5 mm Thickness of flange, t_f = 17.7 mm Root radius, r = 10.2 mm Depth between fillets, d = 407.6 mm Second moment of area about major (*y-y*) axis, *Iy* = 41020 cm^4 Second moment of area about minor (*z-z*) axis, *Iz* = 2089 cm^4	

(continued)

(continued)

Step	Reference	Action/calculation	Conclusion
		Elastic modulus about major (y-y) axis, *Wel,y* = 1770 cm^3 Plastic modulus about major (y-y) axis, *Wpl,y* = 2014 cm^3 Warping constant, I_w = 1.04 dm^6 Torsional constant, I_t = 90.7 cm^4 Area of section, A = 114 cm^2	
4		For ULS, partial factor of safety for both permanent action and variable action selected are 1.35 and 1.5 respectively Uniformly distributed load, w_{ult} $= 1.35G_k + 1.5Q_k$ $= 1.35(10) + 1.5(5)$ **= 21.00 kN/m**	Design load = 21.00 kN/m
		By principle of superposition, V_{Ed} and M_{Ed} for simply supported beam can be determined using equation below: V_{Ed} $= \frac{w_{ult}L}{2} + \frac{R}{2}$ $= \frac{21\times10}{2} + \frac{30}{2}$ **= 120.00 kN**	V_{Ed} = 120.00 kN
		M_{Ed} $= \frac{w_{ult}L^2}{8} + \frac{RL}{4}$ $= \frac{21\times10^2}{8} + \frac{30\times10}{4}$ **= 337.50 kNm**	M_{Ed} = 337.50 kNm
5	Table 5.2	Section classification: i. f_y = 235 N/mm^2 ε = 1 **Class 1** ii. Rolled section, outstand flange: $c = \frac{b-t_w-2r}{2}$ $= \frac{191.9-10.5-2(10.2)}{2}$ = 80.50 mm t_f = 17.7 mm $\frac{C}{t_f} = \frac{80.50}{17.7} = 4.55 < 9\epsilon(= 9)$ **Class 1** iii. Rolled section, web with neutral axis at mid depth: $c^* = d$ = 407.6 mm t_w = 10.5 mm $\frac{c^*}{t_w} = \frac{407.6}{10.5} = 38.82 < 72\epsilon(= 72)$ **Class 1** Therefore, the section is **class 1**	Section class 1
6	SN003b access steel document	Critical buckling resistance can be determined using equation below. For simply supported beam, effective length factor, *K* is taken as 1.0 The addition of secondary beam divides the primary beam into 2 sections with length of 5 m each. The buckling length is hence reduced to 5 m $M_{cr} = \frac{\pi^2 EI_z}{(KL)^2}\sqrt{\left(\frac{I_w}{I_z} + \frac{(KL)^2 GI_t}{\pi^2 EI_z}\right)}$ $= \frac{\pi^2 \times 210 \times 10^9 \times 2089 \times 10^{-8}}{(1.0 \times 5)^2}$ $\times\sqrt{\left(\frac{1.04 \times 10^{-6}}{2089 \times 10^{-8}} + \frac{(1.0 \times 5)^2 \times 81 \times 10^9 \times 90.7 \times 10^{-8}}{\pi^2 \times 210 \times 10^9 \times 2089 \times 10^{-8}}\right)}$ **= 525.88 kNm**	M_{cr} = 525.88 kNm

(continued)

(continued)

Step	Reference	Action/calculation	Conclusion
7	6.3.2.2(1)	For Class 1 section, slenderness for lateral torsional buckling can be determined using equation below: $\bar{\lambda}_{LT} = \sqrt{\frac{W_{pl,y}f_y}{M_{cr}}}$ $= \sqrt{\frac{2014 \times 10^{-6} \times 235 \times 10^6}{525.88 \times 10^3}}$ = **0.95**	$\bar{\lambda}_{LT} = 0.95$
8	Table 6.3 Table 6.4	$\frac{h}{b} = \frac{D}{b} = \frac{463.4}{191.9} = 2.41$ Determine imperfection factor using "General Case" approach: $\frac{h}{b} = 2.41 > 2$ $\alpha_{LT} = 0.34$ $\phi_{LT} = 0.5\left[1 + \alpha_{LT}\left(\bar{\lambda}_{LT} - 0.2\right) + \bar{\lambda}_{LT}^2\right]$ $= 0.5\left[1 + 0.34 \times (0.95 - 0.2) + (0.95)^2\right]$ = **1.08**	$\phi_{LT} = 1.08$
9	6.3.2.2(1)	Lateral torsional buckling reduction factor can be determined using equation below: $\chi_{LT} = \frac{1}{\phi_{LT} + \sqrt{\phi_{LT}^2 - \bar{\lambda}_{LT}^2}}$ $= \frac{1}{1.08 + \sqrt{(1.08)^2 - (0.95)^2}}$ = **0.63**	$\chi_{LT} = 0.63$
10	6.3.2.1(3)	For class 1 section, $M_{b,Rd} = \chi_{LT} W_{pl,y} \frac{f_y}{\gamma_{M1}}$ $= \frac{0.63 \times 2014 \times 10^{-6} \times 235 \times 10^6}{1.0}$ = **298.17 kNm**	$M_{b,Rd}$ = 298.17 kNm
11		$\frac{M_{Ed}}{M_{b,Rd}} = \frac{337.50}{298.17} = \mathbf{1.13 > 1}$ The bending resistance of the section is not adequate	$\frac{M_{Ed}}{M_{b,Rd}} = 1.13$

The section specified is **not suitable** for the situation. Select a larger section and repeat the design.

Step	Reference	Action/calculation	Conclusion
3	Table 3.1	Steel grade = **S235** Assume the thicknesses of web and flange are less than 40 mm: f_y = **235 N/mm²**	f_y = 235 N/mm²
	BS 4 Part 1 2005	Select beam section **533 × 210 × 101** The properties of the section is as follows: Mass per meter = 101 kg/m Depth of section, D = 536.7 mm Width of section, b = 210 mm Thickness of web, t_w = 10.8 mm Thickness of flange, tf = 17.4 mm Root radius, r = 12.7 mm Depth between fillets, d = 476.5 mm Second moment of area about major (y-y) axis, Iy = 61520 cm^4 Second moment of area about minor (z-z) axis, Iz = 2692 cm^4 Elastic modulus about major (y-y) axis, Wel,y = 2292 cm^3	

(continued)

(continued)

Step	Reference	Action/calculation	Conclusion
		Plastic modulus about major (*y*-*y*) axis, Wpl,y $= 2612\ \text{cm}^3$ Warping constant, $I_w = 1.81\ \text{dm}^6$ Torsional constant, $I_t = 101\ \text{cm}^4$ Area of section, $A = 129\ \text{cm}^2$	
4		From previous calculation: V_{Ed} $= \mathbf{120.00\ kN}$	$V_{Ed} = 120.00$ kN
		M_{Ed} $= \mathbf{337.50\ kNm}$	$M_{Ed} = 337.50$ kNm
5	Table 5.2	Section classification: i. $f_y = 235\ \text{N/mm}^2$ $\varepsilon = 1$ **Class 1** ii. Rolled section, outstand flange: $c = \frac{b-t_w-2r}{2}$ $= \frac{210-10.8-2(12.7)}{2}$ $= 86.90$ mm $t_f = 17.4$ mm $\frac{c}{t_f} = \frac{86.90}{17.4} = 4.99 < 9\epsilon(= 9)$ **Class 1** iii. Rolled section, web with neutral axis at mid depth: $c^* = d$ $= 476.5$ mm $t_w = 10.8$ mm $\frac{c^*}{t_w} = \frac{476.5}{10.8} = 44.12 < 72\epsilon(= 72)$ **Class 1** Therefore, the section is **class 1**	Section class 1
6	SN003b access steel document	Critical buckling resistance can be determined using equation below. For simply supported beam, effective length factor, *K* is taken as 1.0: $M_{cr} = \frac{\pi^2 EI_z}{(KL)^2}\sqrt{\left(\frac{I_w}{I_z} + \frac{(KL)^2 GI_t}{\pi^2 EI_z}\right)}$ $= \frac{\pi^2 \times 210 \times 10^9 \times 2692 \times 10^{-8}}{(1.0 \times 5)^2}$ $\times \sqrt{\left(\frac{1.81\times10^{-6}}{2692\times10^{-8}} + \frac{(1.0\times5)^2\times81\times10^9\times101\times10^{-8}}{\pi^2\times210\times10^9\times2692\times10^{-8}}\right)}$ $= \mathbf{719.37\ kNm}$	$M_{cr} = 719.37$ kNm
7	6.3.2.2(1)	For Class 1 section, slenderness for lateral torsional buckling can be determined using equation below: $\bar{\lambda}_{LT} = \sqrt{\frac{W_{pl,y}f_y}{M_{cr}}}$ $= \sqrt{\frac{2612 \times 10^{-6} \times 235 \times 10^6}{719.37 \times 10^3}}$ $= \mathbf{0.92}$	$\bar{\lambda}_{LT} = 0.92$
8	Table 6.3 Table 6.4	$\frac{h}{b} = \frac{D}{b} = \frac{536.7}{210} = 2.56$ Determine imperfection factor using "General Case" approach: $\frac{h}{b} = 2.41 > 2$ $\alpha_{LT} = 0.34$ $\phi_{LT} = 0.5\left[1 + \alpha_{LT}\left(\bar{\lambda}_{LT} - 0.2\right) + \bar{\lambda}_{LT}^2\right]$ $= 0.5\left[1 + 0.34 \times (0.92 - 0.2) + (0.92)^2\right]$ $= \mathbf{1.05}$	$\phi_{LT} = 1.05$

(continued)

(continued)

Step	Reference	Action/calculation	Conclusion
9	6.3.2.2(1)	Lateral torsional buckling reduction factor can be determined using equation below: $\chi_{LT} = \frac{1}{\phi_{LT} + \sqrt{\phi_{LT}^2 - \bar{\lambda}_{LT}^2}}$ $= \frac{1}{1.05 + \sqrt{(1.05)^2 - (0.92)^2}}$ **= 0.64**	$\chi_{LT} = 0.64$
10	6.3.2.1(3)	For Class 1 section, $M_{b,Rd} = \chi_{LT} W_{pl,y} \frac{f_y}{\gamma_{M1}}$ $= \frac{0.64 \times 2692 \times 10^{-6} \times 235 \times 10^6}{1.0}$ = **404.88 kNm**	$M_{b,Rd}$ = 404.88 kNm
11		$\frac{M_{Ed}}{M_{b,Rd}} = \frac{337.50}{404.88} = \mathbf{0.83} < 1$ The bending resistance of the section is adequate	$\frac{M_{Ed}}{M_{b,Rd}} = 0.83$
12	6.2.6(3)	For I beam with load applied on flange, consider the case of rolled I sections with load parallel to web: Shear area, A_v $= A - 2bt_f + (t_w + 2r)t_f$ $= 129 \times 10^2 - 2(210)(17.4) + (10.8 + 2(12.7))(17.4)$ $= 6221.88$ mm^2	
	6.2.6(2)	$V_{pl,Rd} = \frac{A_v(f_y/\sqrt{3})}{\gamma_{M0}}$ $= \frac{6221.88 \times 235}{\sqrt{3}}$ = **844.17 kN**	$V_{pl,Rd}$ = 844.17 kN
13		$\frac{V_{Ed}}{V_{pl,Rd}} = \frac{120.00}{844.17} = \mathbf{0.14} < 1$ The shear resistance is adequate	$\frac{V_{Ed}}{V_{pl,Rd}} = 0.14$
14		For SLS, partial factor of safety or both permanent action and variable action selected is 1.0 Serviceability load, w_{ser} $= 1.0G_k + 1.0Q_k$ $= 1.0(10) + 1.0(5)$ = 15 kN/m By principle of superposition, maximum deflection of the illustrated simply supported beam can be determined using equation below: Maximum deflection, Δ_{max} $= \frac{5wL^4}{384EI} + \frac{PL^3}{48EI}$ $= \frac{5 \times 15 \times 10^3 \times 10^4}{384 \times 210 \times 10^9 \times 61520 \times 10^{-8}} + \frac{30 \times 10^3 \times 10^3}{48 \times 210 \times 10^9 \times 61520 \times 10^{-8}}$ = 0.01996 m = **19.96 mm**	Δ_{max} = 19.96 mm
15	NA2.23	Assume the beam carries plaster of other brittle finishes, Allowable deflection, Δ_{all} $= \frac{L}{360}$ $= \frac{10}{360}$ = 0.02777 m = **27.77 mm**	Δ_{all} = 27.77 mm
16		$\frac{\Delta_{max}}{\Delta_{all}} = \frac{19.96}{27.77} = \mathbf{0.72} < 1$ The deflection is allowable	$\frac{\Delta_{max}}{\Delta_{all}} = 0.72$

(continued)

(continued)

Step	Reference	Action/calculation	Conclusion
17		Check the following ratio: $\frac{M_{Ed}}{M_{b,Rd}} = \frac{337.50}{404.88} = \mathbf{0.83}$ $\frac{V_{Ed}}{V_{pl,Rd}} = \frac{120.00}{844.17} = \mathbf{0.14}$ $\frac{\Delta_{max}}{\Delta_{all}} = \frac{19.96}{27.77} = \mathbf{0.72}$ The values of $\frac{M_{Ed}}{M_{b,Rd}}$ and $\frac{\Delta_{max}}{\Delta_{all}}$ are more than 0.5. Therefore, the beam section 533 × 210 × 101 is considered **optimum**	

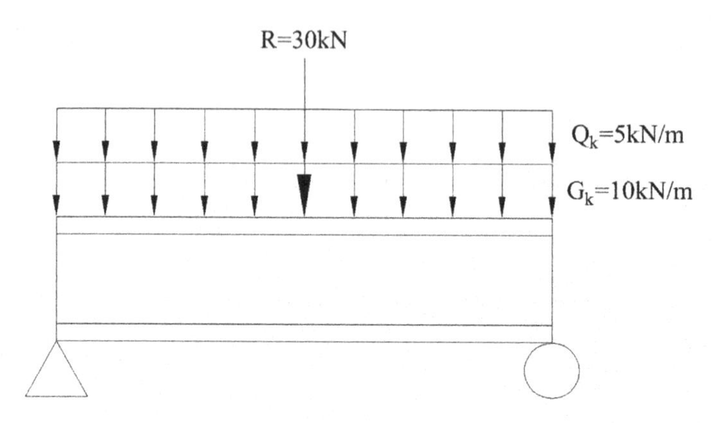

Fig. 2.10 Example 2-5

2.3.4 Example 2-6 Design of a Laterally Unrestrained Beam

Select the optimum section for a cantilever beam subjected to a uniform load (Fig. 2.11). Use steel grade S235 and take the self-weight of the beam into account.

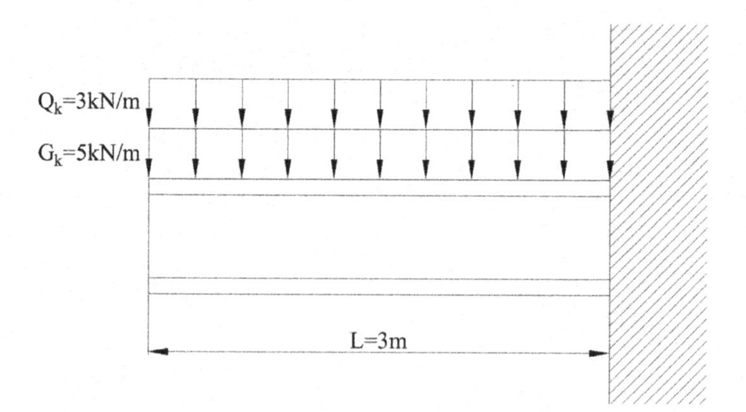

Fig. 2.11 Example 2-6

Step	Reference	Action/calculation	Conclusion
1	References are to BS EN 1993-1-1 unless otherwise stated	From figure, the support condition of beam is **fixed-free**	
2		Permanent action, G_k = **5 kN/m** Variable action, Q_K = **3 kN/m**	
3	Table 3.1	Steel grade = **S235** Assume the thicknesses of web and flange are less than 40 mm: f_y = **235 N/mm²**	$f_y = 235$ N/mm²
	BS 4 Part 1 2005	Randomly choose a beam section for the first trial: Select beam section **254 × 146 × 37** The properties of the section is as follows: Mass per meter = 37 kg/m Depth of section, D = 256 mm Width of section, b = 146.4 mm Thickness of web, t_w = 6.3 mm Thickness of flange, t_f = 10.9 mm Root radius, r = 7.6 mm Depth between fillets, d = 219 mm Second moment of area about major (y-y) axis, Iy = 5537 cm⁴ Second moment of area about minor (z-z) axis, Iz = 571 cm⁴ Elastic modulus about major (y-y) axis, Wel,y = 433 cm³ Plastic modulus about major (y-y) axis, Wpl,y = 483 cm³ Warping constant, I_w = 0.086 dm⁶ Torsional constant, I_t = 15.3 cm⁴ Area of section, A = 47.2 cm²	
4		Self-weight of beam section = 37 kg/m × 9.81 N/kg = **0.36 kN/m** For ULS, partial factor of safety for both permanent action and variable action selected are 1.35 and 1.5 respectively Uniformly distributed load, w_{ult} $= 1.35G_k + 1.5Q_k$ = 1.35(5 + 0.36) + 1.5(3) = **11.74 kN/m**	Design load = 11.74 kN/m
		For cantilever, V_{Ed} and M_{Ed} can be determined using equation below: V_{Ed} $= w_{ult}L$ = 11.74 × 3 = **35.22 kN**	V_{Ed} = 35.22 kN
		M_{Ed} $= \frac{w_{ult}L^2}{2}$ $= \frac{11.74 \times 3^2}{2}$ = **52.83 kNm**	M_{Ed} = 52.83 kNm
5	Table 5.2	Section classification: i. f_y = 235 N/mm² $\varepsilon = 1$ **Class 1**	Section class 1

(continued)

(continued)

Step	Reference	Action/calculation	Conclusion
		ii. Rolled section, outstand flange: $c = \frac{b - t_w - 2r}{2}$ $= \frac{146.4 - 6.3 - 2(7.6)}{2}$ $= 62.45$ mm $t_f = 10.9$ mm $\frac{c}{t_f} = \frac{62.45}{10.9} = 5.73 < 9\epsilon(= 9)$ **Class 1** iii. Rolled section, web with neutral axis at mid depth: $c^* = d$ $= 219$ mm $t_w = 6.3$ mm $\frac{c^*}{t_w} = \frac{219}{6.3} = 34.76 < 72\epsilon(= 72)$ **Class 1** Therefore, the section is **class 1**	
6	SN003b access steel document	Critical buckling resistance can be determined using equation below. For cantilever, effective length factor, K is taken as 2.0: $M_{cr} = \frac{\pi^2 EI_z}{(KL)^2}\sqrt{\left(\frac{I_w}{I_z} + \frac{(KL)^2 GI_t}{\pi^2 EI_z}\right)}$ $= \frac{\pi^2 \times 210 \times 10^9 \times 571 \times 10^{-8}}{(2.0 \times 3)^2}$ $\times \sqrt{\left(\frac{0.086 \times 10^{-6}}{571 \times 10^{-8}} + \frac{(2.0 \times 3)^2 \times 81 \times 10^9 \times 15.3 \times 10^{-8}}{\pi^2 \times 210 \times 10^9 \times 571 \times 10^{-8}}\right)}$ $=$ **75.51 kNm**	$M_{cr} = 75.51$ kNm
7	6.3.2.2(1)	For Class 1 section, slenderness for lateral torsional buckling can be determined using equation below: $\bar{\lambda}_{LT} = \sqrt{\frac{W_{pl,y} f_y}{M_{cr}}}$ $= \sqrt{\frac{483 \times 10^{-6} \times 235 \times 10^6}{75.51 \times 10^3}}$ $=$ **1.23**	$\bar{\lambda}_{LT} = 1.23$
8	Table 6.3 Table 6.4	$\frac{h}{b} = \frac{D}{b} = \frac{256}{146.4} = 1.75$ Determine imperfection factor using "Rolled Section" approach: $\frac{h}{b} = 1.75 < 2$ Using "Rolled Section" approach, $\alpha_{LT} = 0.34$ $\phi_{LT} = 0.5\left[1 + \alpha_{LT}\left(\bar{\lambda}_{LT} - 0.4\right) + 0.75\bar{\lambda}_{LT}^2\right]$ $= 0.5\left[1 + 0.34 \times (1.23 - 0.4) + 0.75 \times (1.23)^2\right]$ $=$ **1.21**	$\phi_{LT} = 1.21$
9	6.3.2.2(1)	Lateral torsional buckling reduction factor can be determined using equation below: $\chi_{LT} = \frac{1}{\phi_{LT} + \sqrt{\phi_{LT}^2 - 0.75\bar{\lambda}_{LT}^2}}$ $= \frac{1}{1.21 + \sqrt{(1.21)^2 - 0.75 \times (1.23)^2}}$ $= 0.56$ $\frac{1}{\bar{\lambda}_{Lt}^2} = \frac{1}{1.23^2} = 0.66 > \chi_{LT}(= 0.56)$	$\chi_{LT,mod} = 0.61$

(continued)

(continued)

Step	Reference	Action/calculation	Conclusion
		Bending moment diagram for the beam is shown as below: 0 52.83kNm The moment distribution is compared with the tabulated pattern. K_C is taken as $\frac{1}{1.33-0.33\psi}$ Ratio of moment at two ends should between −1 and 1. So, the numerator and denominator should be arranged accordingly to make the result falls within the range: $\psi = \frac{0}{52.83} = 0$ $K_C = \frac{1}{1.33-0.33\times 0} = 0.75$ $f = 1 - 0.5(1 - K_C)\left[1 - 2(\bar{\lambda}_{LT} - 0.8)^2\right]$ $= 1 - 0.5(1 - 0.75)\left[1 - 2(1.23 - 0.8)^2\right]$ $= 0.92$ Lateral torsional buckling reduction factor can be determined using equation below: $\chi_{LT,mod} = \frac{\chi_{LT}}{f} = \frac{0.56}{0.92} = 0.61$	
10	6.3.2.1(3)	For Class 1 section, $M_{b,Rd} = \chi_{LT} W_{pl,y} \frac{f_y}{\gamma_{M1}}$ $= \frac{0.61 \times 483 \times 10^{-6} \times 235 \times 10^6}{1.0}$ **= 69.24 kNm**	$M_{b,Rd}$ = 69.24 kNm
11		$\frac{M_E}{M_{b,Rd}} = \frac{52.83}{69.24} = \mathbf{0.76} < 1$ The bending resistance of the section is adequate	$\frac{M_{Ed}}{M_{b,Rd}} = 0.76$
12	6.2.6(3)	For I beam with load applied on flange, consider the case of rolled I sections with load parallel to web: Shear area, A_v $= A - 2bt_f + (t_w + 2r)t_f$ $= 47.2 \times 10^2 - 2(146.4)(10.9) + (6.3 + 2(7.6))(10.9)$ $= 1762.83\ mm^2$	
	6.2.6(2)	$V_{pl,Rd} = \frac{A_v(f_y/\sqrt{3})}{\gamma_{M0}}$ $= \frac{1762.83\times 235}{\sqrt{3}}$ $=$ **239.18 kN**	$V_{pl,Rd}$ = 239.18 kN
13		$\frac{V_{Ed}}{V_{pl,Rd}} = \frac{35.22}{239.18} = \mathbf{0.15} < 1$ The shear resistance is adequate	$\frac{V_{Ed}}{V_{pl,Rd}} = 0.15$
14		For SLS, partial factor of safety or both permanent action and variable action selected is 1.0. Serviceability load, w_{ser} $= 1.0G_k + 1.0Q_k$ $= 1.0(5.36) + 1.0(3)$ $= 8.36kN/m$ For cantilever, maximum deflection can be determined using equation below: $= \frac{wL^4}{8EI}$ $= \frac{8.36\times 10^3 \times 3^4}{8\times 210\times 10^9 \times 5537\times 10^{-8}}$ $= 7.28 \times 10^{-3} m$ $=$ **7.28 mm**	Δ_{max} = 7.28 mm

(continued)

(continued)

Step	Reference	Action/calculation	Conclusion
15	NA2.23	For cantilever beam, Allowable deflection, Δ_{all} $= \frac{L}{180}$ $= \frac{3}{180}$ = 0.01667 m = **16.67 mm**	Δ_{all} = 16.67 mm
16		$\frac{\Delta_{max}}{\Delta_{all}} = \frac{7.28}{16.67} = \mathbf{0.44} < 1$ The deflection is allowable	$\frac{\Delta_{max}}{\Delta_{all}} = 0.44$
17		Check the following ratio: $\frac{M_{Ed}}{M_{b,Rd}} = \frac{52.83}{69.24} = \mathbf{0.76}$ $\frac{V_{Ed}}{V_{pl,Rd}} = \frac{35.22}{239.18} = \mathbf{0.15}$ $\frac{\Delta_{max}}{\Delta_{all}} = \frac{7.28}{16.67} = \mathbf{0.44}$ The values of $\frac{M_{Ed}}{M_{b,Rd}}$ is more than 0.5. Therefore, the beam section 254 × 146 × 37 is **adequate**. However, a smaller beam section may be selected	

Step 3 is repeated to using a smaller section.

Step	Reference	Action/calculation	Conclusion
3	Table 3.1	Steel grade = **S235** Assume the thicknesses of web and flange are less than 40 mm: f_y = **235 N/mm²**	f_y = 235 N/mm²
	BS 4 Part 1 2005	Select beam section **254 × 146 × 31** The properties of the section is as follows: Mass per meter = 31.1 kg/m Depth of section, D = 251.4 mm Width of section, b = 146.1 mm Thickness of web, t_w = 6.0 mm Thickness of flange, t_f = 8.6 mm Root radius, r = 7.6 mm Depth between fillets, d = 219.0 mm Second moment of area about major (y-y) axis, Iy = 4413 cm⁴ Second moment of area about minor (z-z) axis, Iz = 448 cm⁴ Elastic modulus about major (y-y) axis, Wel,y = 351 cm³ Plastic modulus about major (y-y) axis, Wpl,y = 393 cm³ Warping constant, I_w = 0.066 dm⁶ Torsional constant, I_t = 8.55 cm⁴ Area of section, A = 39.7 cm²	
4		Self-weight of beam section = 31.1 kg/m × 9.81 N/kg = **0.31 kN/m** For ULS, partial factor of safety for both permanent action and variable action selected are 1.35 and 1.5 respectively. Uniformly distributed load, w_{ult} $= 1.35G_k + 1.5Q_k$ = 1.35(5 + 0.31) + 1.5(3) = **11.67 kN/m**	Design load = 11.67 kN/m

(continued)

(continued)

Step	Reference	Action/calculation	Conclusion
		For cantilever, V_{Ed} and M_{Ed} can be determined using equation below: V_{Ed} $= w_{ult}L$ $= 11.67 \times 3$ $= \mathbf{35.01\ kN}$	V_{Ed} = 35.01 kN
		M_{Ed} $= \frac{w_{ult}L^2}{2}$ $= \frac{11.67 \times 3^2}{2}$ $= \mathbf{52.52\ kNm}$	M_{Ed} = 52.52 kNm
5	Table 5.2	Section classification: i. f_y = 235 N/mm^2 $\varepsilon = 1$ **Class 1** ii. Rolled section, outstand flange: $c = \frac{b - t_w - 2r}{2}$ $= \frac{146.1 - 6.0 - 2(7.6)}{2}$ = 62.45 mm t_f = 8.6 mm $\frac{c}{t_f} = \frac{62.45}{8.6} = 7.26 < 9\epsilon (= 9)$ **Class 1** iii. Rolled section, web with neutral axis at mid depth: $c^* = d$ = 219.0 mm t_w = 6.0 mm $\frac{c^*}{t_w} = \frac{219.0}{6.0} = 36.50 < 72\epsilon (= 72)$ **Class 1** Therefore, the section is **class 1**	Section class 1
6	SN003b access steel document	Critical buckling resistance can be determined using equation below. For cantilever, effective length factor, K is taken as 2.0: $M_{cr} = \frac{\pi^2 EI_z}{(KL)^2}\sqrt{\left(\frac{I_w}{I_z} + \frac{(KL)^2 GI_t}{\pi^2 EI_z}\right)}$ $= \frac{\pi^2 \times 210 \times 10^9 \times 448 \times 10^{-8}}{(2.0 \times 3)^2}$ $\times \sqrt{\left(\frac{0.066 \times 10^{-6}}{448 \times 10^{-8}} + \frac{(2.0 \times 3)^2 \times 81 \times 10^9 \times 8.55 \times 10^{-8}}{\pi^2 \times 210 \times 10^9 \times 448 \times 10^{-8}}\right)}$ $= \mathbf{52.60\ kNm}$	M_{cr} = 52.60 kNm
7	6.3.2.2(1)	For Class 1 section, slenderness for lateral torsional buckling can be determined using equation below: $\bar{\lambda}_{LT} = \sqrt{\frac{W_{pl,y} f_y}{M_{cr}}}$ $= \sqrt{\frac{393 \times 10^{-6} \times 235 \times 10^6}{52.60 \times 10^3}}$ = **1.33**	$\bar{\lambda}_{LT} = 1.33$

(continued)

(continued)

Step	Reference	Action/calculation	Conclusion
8	Table 6.3 Table 6.4	$\frac{h}{b} = \frac{D}{b} = \frac{251.4}{146.1} = 1.7$ Determine imperfection factor using "Rolled Section" approach: $\frac{h}{b} = 1.7 < 2$ Using "Rolled Section" approach, $\alpha_{LT} = 0.34$ $\phi_{LT} = 0.5\left[1 + \alpha_{LT}\left(\bar{\lambda}_{LT} - 0.4\right) + 0.75\bar{\lambda}_{LT}^{2}\right]$ $= 0.5\left[1 + 0.34 \times (1.33 - 0.4) + 0.75 \times (1.33)^{2}\right]$ $= \mathbf{1.32}$	$\phi_{LT} = 1.32$
9	6.3.2.2(1)	Lateral torsional buckling reduction factor can be determined using equation below: $\chi_{LT} = \frac{1}{\phi_{LT} + \sqrt{\phi_{LT}^{2} - 0.75\bar{\lambda}_{LT}^{2}}}$ $= \frac{1}{1.32 + \sqrt{(1.32)^{2} - 0.75 \times (1.33)^{2}}}$ $= 0.51$ $\frac{1}{\bar{\lambda}_{LT}^{2}} = \frac{1}{1.33^{2}} = 0.56 > \chi_{LT}(= 0.51)$ Bending moment diagram for the beam is shown as below: 0 52.52kNm The moment distribution is compared with the tabulated pattern. $K_C$is taken as $\frac{1}{1.33 - 0.33\psi}$ Ratio of moment at two ends should between −1 to 1. So, the numerator and denominator should be arranged accordingly to make the result falls within the range: $\psi = \frac{0}{52.52} = 0$ $K_C = \frac{1}{1.33 - 0.33 \times 0} = 0.75$ $f = 1 - 0.5(1 - K_C)\left[1 - 2\left(\bar{\lambda}_{LT} - 0.8\right)^{2}\right]$ $= 1 - 0.5(1 - 0.75)\left[1 - 2(1.33 - 0.8)^{2}\right]$ $= 0.95$ Lateral torsional buckling reduction factor can be determined using equation below: $\chi_{LT,mod} = \frac{\chi_{LT}}{f} = \frac{0.51}{0.95} = \mathbf{0.54}$	$\chi_{LT,mod} = 0.54$
10	6.3.2.1(3)	For Class 1 section, $M_{b,Rd} = \chi_{LT} W_{pl,y} \frac{f_y}{\gamma_{M1}}$ $= \frac{0.54 \times 393 \times 10^{-6} \times 235 \times 10^{6}}{1.0}$ $= \mathbf{49.87\ kNm}$	$M_{b,Rd} = 49.87$ kNm
11		$\frac{M_E}{M_{b,Rd}} = \frac{52.52}{49.87} = \mathbf{1.05} > 1$ The bending resistance of the section is not adequate The beam section 254 × 146 × 31 is found unsuitable. Therefore, the beam section selected for first trial, 254 × 146 × 37 is concluded as an **optimum** section	$\frac{M_{Ed}}{M_{b,Rd}} = 1.05$

2.4 Exercise: Beam Design

2-1 A secondary beam is connected to the primary beam by shear connection (Fig. 2.12). Select the optimum section for the primary beam. Use steel grade S235. Assume the primary beam is laterally unrestrained and sits on 100 mm bearings at each end. Ignore the self-weight of the beam.

2-2 Check the suitability of a 305 × 165 × 46 section for the beam shown in Fig. 2.13. Use steel grade S275 and assume the beam is laterally unrestrained. Take the self-weight of the beam into account. Compare the bending moment resistances obtained when rolled section and the general case approaches are used.

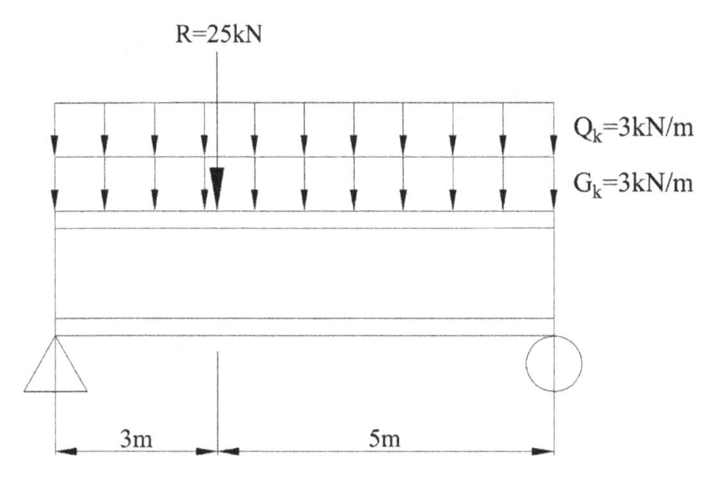

Fig. 2.12 Question 2-1

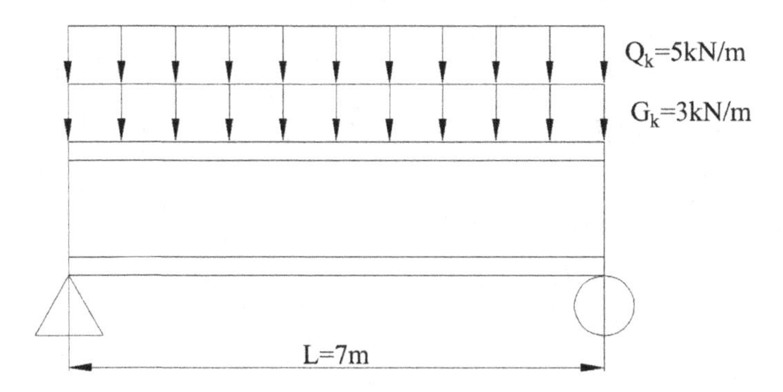

Fig. 2.13 Question 2-2

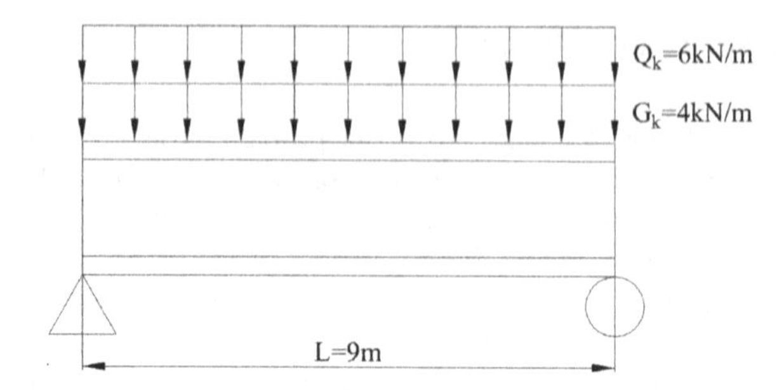

Fig. 2.14 Question 2-3

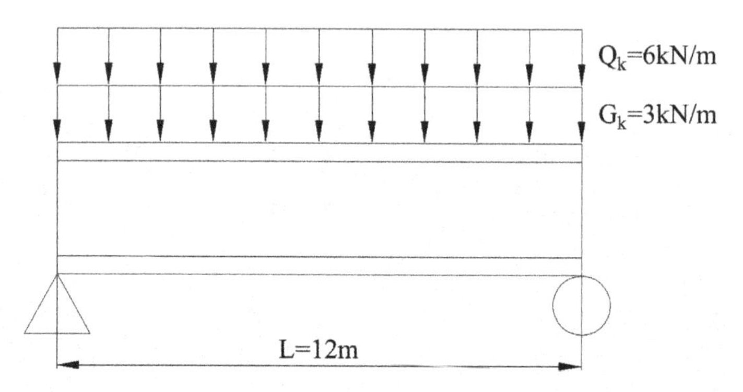

Fig. 2.15 Question 2-4

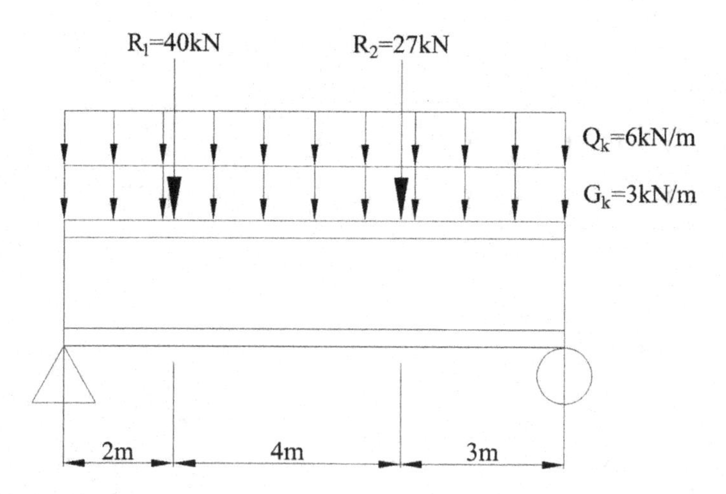

Fig. 2.16 Question 2-5

2-3 Select the optimum section for the beam in Fig. 2.14. Use steel grade S235 and assume the beam is laterally restrained. Consider the self-weight of the beam.

2-4 Select the optimum section for the beam in Fig. 2.15. Use steel grade S235 and assume the beam is laterally restrained. Consider the self-weight of the beam.

2-5 Select the optimum section for the beam in Fig. 2.16. Use steel grade S275. Assume the primary beam is laterally unrestrained and sits on 100 mm bearings at each end. Ignore the self-weight of the beam.

Chapter 3
Column Design

3.1 Introduction

Column is a structural member that supports beams and slabs by carrying their loads down to the foundation. The direction of its load is along the longitudinal axis (*x*-*x*). Thus, column is primarily a compression member (Fig. 3.1).

Other than an axial load, a column may also be subjected to a bending moment. This bending moment is usually due to the eccentricity of the reaction force from the beam or the slab.

A column can be categorized either as short or slender based on the slenderness ratio. Slenderness ratio is the ratio of column length to its cross-sectional effective width. A high slenderness ratio indicates a slender column. A short column usually fails by crushing, whereas a slender column usually fails by buckling (Fig. 3.2).

In EC3, a column can be designed using a simplified approach. This approach, however, is only applicable to simple construction. The beam–column connection must be pinned, and the bending moment resulting from the eccentricity of the beam–column connection should be insignificant.

F. Hejazi and T. K. Chun, *Steel Structures Design Based on Eurocode 3*,
https://doi.org/10.1007/978-981-10-8836-0_3

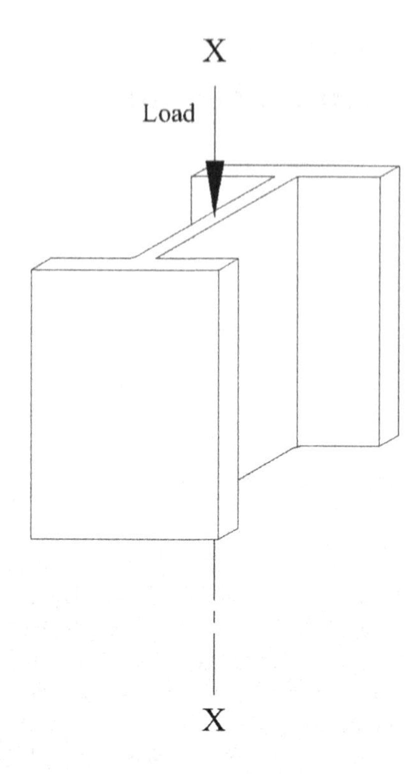

Fig. 3.1 Column and its loading

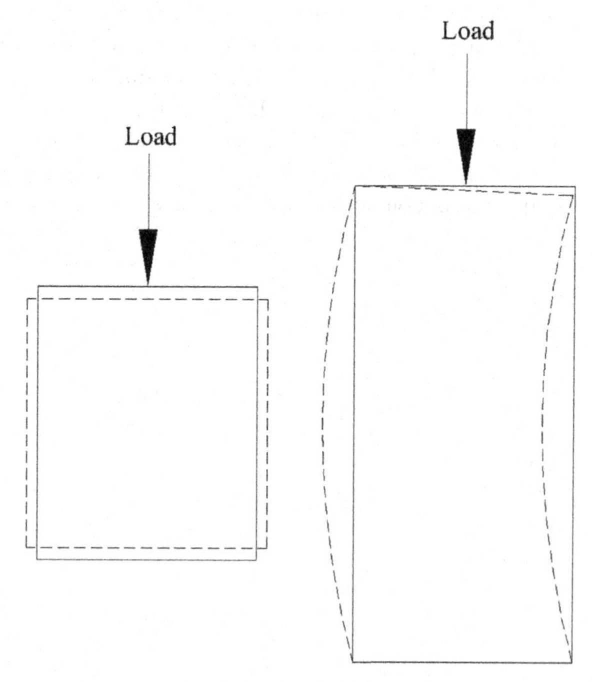

Fig. 3.2 Failure modes of columns

Table 3.1 Nominal values of yield strength f_y and ultimate tensile strength f_u of hot-rolled structural steel (BS EN 1993-1-1:2005 Table 3.1)

Standard and Steel Grade (To BS EN 10025-2)	Nominal Thickness of element, t (mm)			
	$t \leq 40$ mm		40 mm $< t \leq 80$ mm	
	f_y(N/mm^2)	f_u(N/mm^2)	f_y(N/mm^2)	f_u(N/mm^2)
S235	235	360	215	360
S275	275	430	255	410
S355	355	490	335	470
S450	440	550	410	550

3.2 Design Procedure for a Column

The design procedure for a column is as follows:

1. Determine the support condition (i.e., pin, roller, or fixed at the base of the column).
2. Determine the reaction of the beams.
3. Choose the steel grade (refer to Table 3.1). Refer to BS 4 Part 1 2005 to choose the column section for use in construction. A table for the universal section commonly used for columns and their corresponding properties is provided in Appendix A.3.
4. Determine the design axial load and the design bending moments about the *y-y* and *z-z* axes. Design axial load is the summation of the total reaction (the design shear force of the beam) at the beam–column connection and the load applied to the column. The design bending moment about the *y-y* and the *z-z* axes is the moment induced by the eccentricity of the beam–column connection. In other words, ensuring that the shear force acting on the beam will act on the centroid of the column is difficult, and consequently, column bending will occur because of such eccentricity. The bending moment about the *y-y* axis is induced by the beam connected to the column flange, and the bending moment about the *z-z* axis is induced by the beam connected to the column web. The point at which shear force acts on the beam depends on the size of the bearing where the edges of the beam stand. Given that the moments induced by the opposite sides of the flange and the web about the same axis are in opposite directions, these moments will counter each other.
   According to the SN005a-EN-EU Access Steel document, the beam reaction is assumed to act at 100 mm from the face of the column. Therefore, if the bearing size is not specified, the beam reaction can be assumed to be 100 mm.

$$N_{Ed} = \sum_{i=1}^{n} V_{Ed,i} + load\ on\ column \tag{3.1}$$

where V_{Ed} is reaction of beams obtained from Step 2

$$M_{y,Ed} = Shear\ difference\ in\ y\text{-}y \times \left(\frac{D}{2} + bearing\ size\right) \quad (3.2)$$

where D is depth of column section by referring to Appendix A.3

$$M_{z,Ed} = Shear\ difference\ in\ z\text{-}z \times \left(\frac{t_w}{2} + bearing\ size\right) \quad (3.3)$$

where t_w is thickness of web of column section by referring to Appendix A.3

5. Classi*fy* the column section. To carry out the classification, check only under the criteria "outstand flange for rolled sections" and "web subject to compression, rolled sections" (Table 3.2).

Table 3.2 Maximum width-to-thickness ratio of the compression element (BS EN 1993-1-1:2005 Table 5.2)

Type of element	Class of element		
	Class 1	Class 2	Class 3
Outstand flange for rolled section	$c/t_f \leq 9\,\varepsilon$	$c/t_f \leq 10\,\varepsilon$	$c/t_f \leq 14\,\varepsilon$
Web with neutral axis at mid depth, rolled sections	$c^*/t_w \leq 72\,\varepsilon$	$c^*/t_w \leq 83\,\varepsilon$	$c^*/t_w \leq 124\,\varepsilon$
Web subject to compression, rolled sections	$c^*/t_w \leq 33\,\varepsilon$	$c^*/t_w \leq 38\,\varepsilon$	$c^*/t_w \leq 42\,\varepsilon$
f_y	235	275	355
ε	1	0.92	0.81

Where t_f is thickness of flange by referring to Appendix A.3
t_w is thickness of web by referring to Appendix A.3
$c^* = d$ by referring to Appendix A.2
$c = (b - t_w - 2r)/2$

6. Determine the non-dimensional slenderness $\bar{\lambda}$. When the support conditions at the base of the column about the *y-y* and *z-z* axes are different, the non-dimensional slenderness for both the *y-y* and *z-z* axes should be considered. Otherwise, consider only the minor axis.

$$\bar{\lambda} = \frac{KL}{i} \times \frac{1}{\pi}\left(\sqrt{\frac{f_y}{E}}\right) \quad (3.4)$$

where K is effective length factor obtained from Step 6 (Table 3.3)

Table 3.3 Values of the effective length factor K for different support conditions (BS5950: Part 1 4.7.10)

Support condition	Effective length factor, K
Fixed-Fixed	0.7
Fixed-Pinned	0.85
Pinned-Pinned	1.0
Fixed-Free	2.0

L is length of column
i is radius of gyration by referring to Appendix A.3
f_y is yield strength of steel obtained from Step 3 (Table 3.1)
E is modulus of elasticity of steel = 210×10^9 N/m^2

(BS EN 1993-1-1:2005 6.3.1.3(1))

7. Determine Φ. Consider only the minor axis to determine the imperfection factors.

$$\phi = 0.5\left[1 + \alpha(\bar{\lambda} - 0.2) + \bar{\lambda}^2\right] \tag{3.5}$$

where h is depth of section by referring to Appendix A.3
b is width of section by referring to Appendix A.3
t_f is thickness of flange by referring to Appendix A.3
α is imperfection factor obtained from Step 7 (Table 3.4)
$\bar{\lambda}$ is non-dimensional slenderness obtained from Step 6 (Eq. 3.4)

(BS EN 1993-1-1:2005 6.3.1.2(1))

Table 3.4 Values of the imperfection factor α for different section geometries (BS EN 1993-1-1:2005 Tables 6.1 and 6.2)

Limits		Buckling about axis	Imperfection factor, α
$\frac{h}{b} \geq 1.2$	$t_f \leq 40$ mm	y-y	0.21
		z-z	0.34
	40 mm $< t_f \leq 100$ mm	y-y	0.34
		z-z	0.49
$\frac{h}{b} \leq 1.2$	$t_f \leq 100$ mm	y-y	0.34
		z-z	0.49
	$t_f > 100$ mm	y-y	0.76
		z-z	0.76

8. Determine the reduction factor χ.

$$\chi = \frac{1}{\phi + \sqrt{\phi^2 - \bar{\lambda}^2}} \leq 1.0 \tag{3.6}$$

where ϕ is obtained from Step 7 (Eq. 3.5)
$\bar{\lambda}$ is non-dimensional slenderness obtained from Step 6 (Eq. 3.4)

(BS EN 1993-1-1:2005 6.3.1.2(1))

9. Determine the buckling resistance of the column.

$$N_{b,d} = \begin{cases} \frac{\chi A f_y}{\gamma_{M1}}, \textit{Class 1, 2 and 3 sections} \\ \frac{\chi A_{eff} f_y}{\gamma_{M1}}, \textit{Class 4 sections} \end{cases} \tag{3.7}$$

where A is area of section by referring to Appendix A.3
A_{eff} is effective area of section
f_y is yield strength of steel obtained from Step 3 (Table 3.1)

(BS EN 1993-1-1:2005 6.3.1.1(3))

10. Compare the design compression force and buckling resistance of the column. If the design compression force exceeds the design buckling resistance of the column, repeat Step 3 to choose a better section. Otherwise, proceed to Step 11.
11. Determine the critical buckling moment. The support condition influences the effective length of the member subjected to buckling (refer to Appendix A.3 for the section properties of column sections and Table 3.3 for the values of K).

$$M_{cr} = \frac{\pi^2 EI_z}{(KL)^2}\sqrt{\left(\frac{I_w}{I_z} + \frac{(KL)^2 GI_t}{\pi^2 EI_z}\right)} \tag{3.8}$$

where E is modulus of elasticity of steel = 210×10^9 N/m^2
I_z is second moment of area about z-z axis by referring to Appendix A.3
K is effective length factor obtained from Step 6 (Table 3.3)
L is length of column
I_w is warping constant by referring to Appendix A.3
G is shear modulus of steel = 81×10^9 N/m^2
I_t is torsional constant by referring to Appendix A.3

(SN003b Access Steel document)

12. Determine the slenderness for lateral-torsional buckling $\bar{\lambda}_{LT}$.

$$\bar{\lambda}_{LT} = \begin{cases} \sqrt{\frac{W_{pl,y} f_y}{M_{cr}}}, \textit{Class 1 and 2 sections} \\ \sqrt{\frac{W_{el,y} f_y}{M_{cr}}}, \textit{Class 3 sections} \\ \sqrt{\frac{W_{eff,y} f_y}{M_{cr}}}, \textit{Class 4 sections} \end{cases} \tag{3.9}$$

where:

$W_{pl,y}$ is plastic section modulus about y-y axis by referring to Appendix A.3
$W_{el,y}$ is elastic section modulus about y-y axis by referring to Appendix A.3
$W_{eff,y}$ is effective section modulus about y-y axis
f_y is yield strength of steel obtained from Step 3 (Table 3.1)
M_{cr} is critical buckling moment obtained from Step 11 (Eq. 3.8)

(BS EN 1993-1-1:2005 6.3.2.2(1))

Table 3.5 Values of the imperfection factor α_{LT} for different approaches (BS EN 1993-1-1:2005 Tables 6.3 and 6.4)

Limit	α_{LT}
$h/b \leq 2$	0.21
$h/b > 2$	0.34

Where h is depth of section by referring to Appendix A.3
b is width of section by referring to Appendix A.3

13. Determine the imperfection factors for lateral-torsional buckling, α_{LT} and ϕ_{LT}.

$$\phi_{LT} = 0.5\left[1 + \alpha_{LT}\left(\bar{\lambda}_{LT} - 0.2\right) + \bar{\lambda}_{LT}^2\right] \quad (3.10)$$

where α_{LT} is imperfection factor obtained from Step 13 (Table 3.5)
$\bar{\lambda}_{LT}$ is slenderness for lateral torsional buckling obtained from Step 12 (Eq. 3.9)

(BS EN 1993-1-1:2005 6.3.2.2(1))

14. Determine the lateral torsional buckling reduction factor χ_{LT}.

$$\chi_{LT} = \frac{1}{\phi_{LT} + \sqrt{\phi_{LT}^2 - \bar{\lambda}_{LT}^2}} \quad (3.11)$$

where ϕ_{LT} is obtained from Step 13 (Eq. 3.10)
$\bar{\lambda}_{LT}$ is slenderness for lateral torsional buckling obtained from Step 12 (Eq. 3.9)

(BS EN 1993-1-1:2005 6.3.2.2(1))

15. Determine the buckling moment resistance.

$$M_{b,Rd} = \begin{cases} \chi_{LT} W_{pl,y} \frac{f_y}{\gamma_{M1}}, \mathit{Class\ 1\ and\ 2\ sections} \\ \chi_{LT} W_{el,y} \frac{f_y}{\gamma_{M1}}, \mathit{Class\ 3\ sections} \\ \chi_{LT} W_{eff,y} \frac{f_y}{\gamma_{M1}}, \mathit{Class\ 4\ sections} \end{cases} \quad (3.12)$$

where

$W_{pl,y}$ is plastic section modulus about y-y axis by referring to Appendix A.3
$W_{el,y}$ is elastic section modulus about y-y axis by referring to Appendix A.3
$W_{eff,y}$ is effective section modulus about y-y axis
f_y is yield strength of steel obtained from Step 3 (Table 3.1)
χ_{LT} is lateral torsional buckling reduction factor obtained from Step 14 (Eq. 3.11)

(BS EN 1993-1-1:2005 6.3.2.1(3))

16. Compare the design bending moment of the structure and the buckling moment resistance of the section. If the buckling moment resistance of the structure is insufficient, repeat Step 3 to choose a better section. Otherwise, proceed to Step 17.

17. Determine the bending moment resistance about the z-z axis.

$$M_{z,d} = \begin{cases} \frac{W_{pl,z} f_y}{\gamma_{M1}}, \textit{Class 1 and 2 sections} \\ \frac{W_{el,z} f_y}{\gamma_{M1}}, \textit{Class 3 sections} \end{cases} \tag{3.13}$$

where

$W_{pl,z}$ is plastic section modulus about z-z axis by referring to Appendix A.3
$W_{el,z}$ is elastic section modulus about z-z axis by referring to Appendix A.3
f_y is yield strength of steel obtained from Step 3 (Table 3.1)

(BS EN 1993-1-1:2005 6.2.5(2))

18. Refer to the SN048a-EN-GB Access Steel document to determine the combined ratio of the design load to the resistance of the column. If the ratio is greater than 1, repeat Step 3 to choose a better section. Otherwise, proceed to Step 19.

$$\frac{N_{Ed}}{N_{b,Rd}} + \frac{M_{y,Ed}}{M_{b,Rd}} + 1.5\frac{M_{z,Ed}}{M_{z,Rd}} \leq 1.0 \tag{3.14}$$

where $\frac{N_{Ed}}{N_{b,Rd}}$ is ratio obtained from Step 10
$\frac{M_{y,Ed}}{M_{b,Rd}}$ is ratio obtained from Step 16
$M_{z,Ed}$ is design bending moment about z-z axis of column obtained from Step 4 (Eq. 3.3)
$M_{z.Rd}$ is bending moment resistance about the z-z axis obtained from Step 17 (Eq. 3.13)

(SN048b-EN-GB)

19. Check whether the section is an overdesign by checking the ratio obtained in Step 18. If the ratio is less than 0.5, repeat Step 3 and choose a smaller section to ensure optimum design.

3.2.1 Design Flowchart for a Column

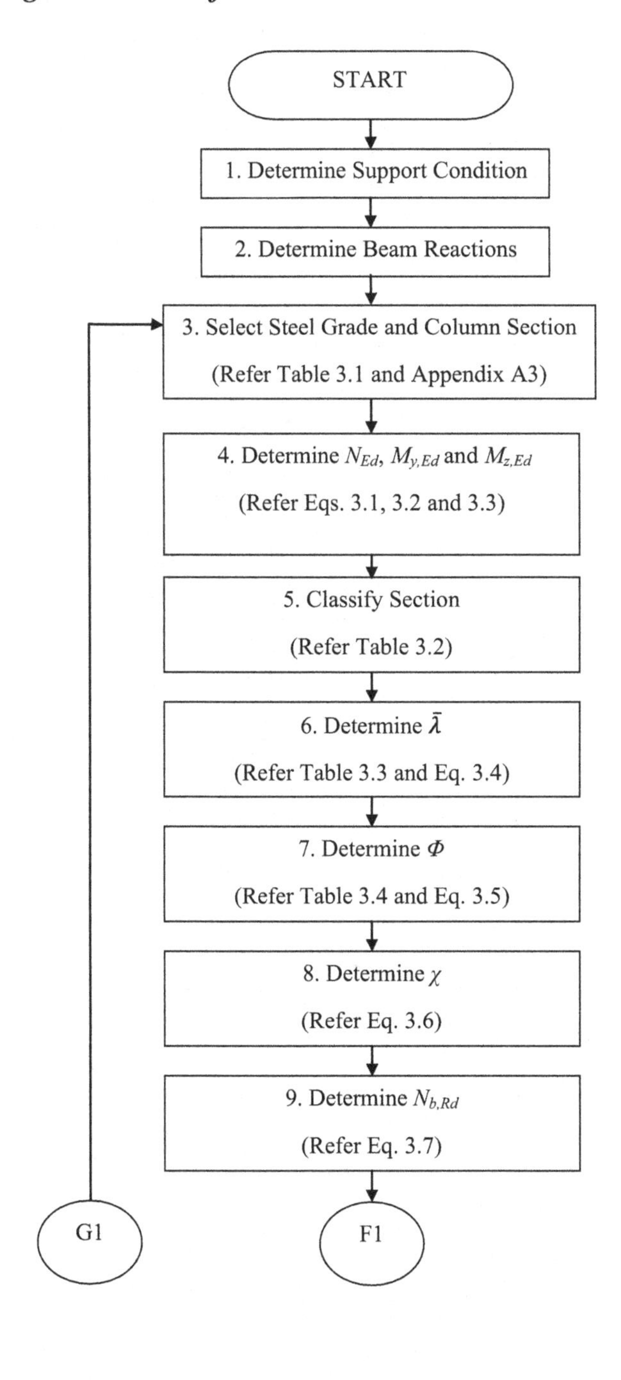

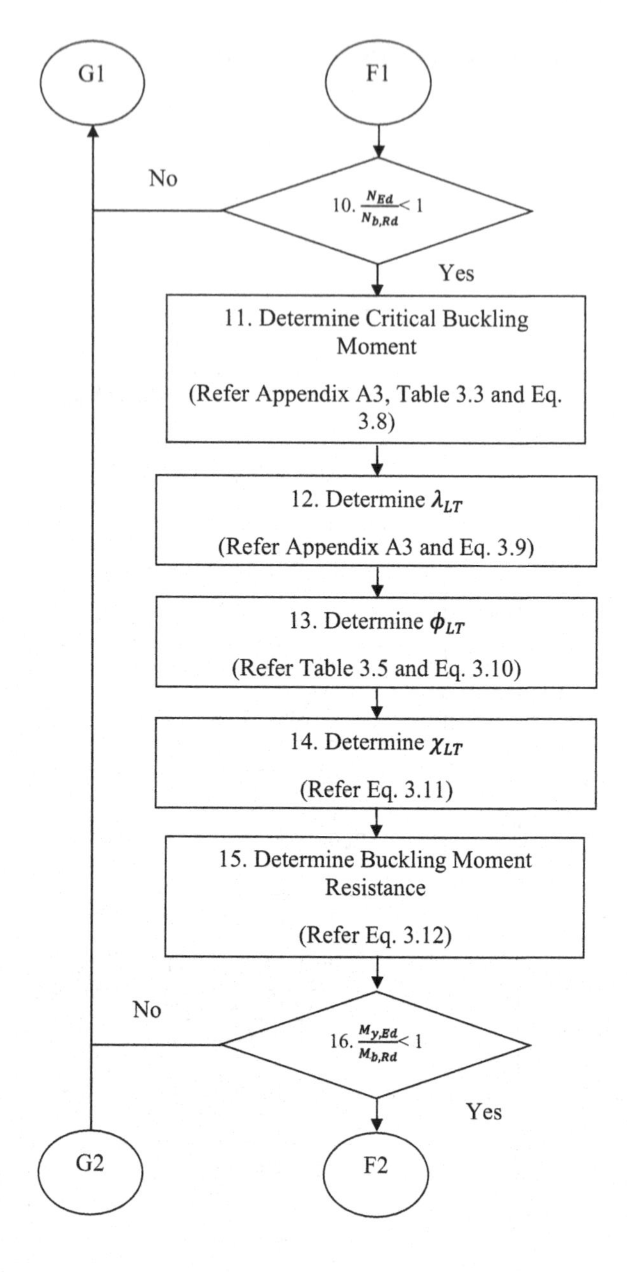
G1
F1
No
10. $\frac{N_{Ed}}{N_{b,Rd}} < 1$
Yes
11. Determine Critical Buckling Moment
(Refer Appendix A3, Table 3.3 and Eq. 3.8)
12. Determine λ_{LT}
(Refer Appendix A3 and Eq. 3.9)
13. Determine ϕ_{LT}
(Refer Table 3.5 and Eq. 3.10)
14. Determine χ_{LT}
(Refer Eq. 3.11)
15. Determine Buckling Moment Resistance
(Refer Eq. 3.12)
No
16. $\frac{M_{y,Ed}}{M_{b,Rd}} < 1$
Yes
G2
F2

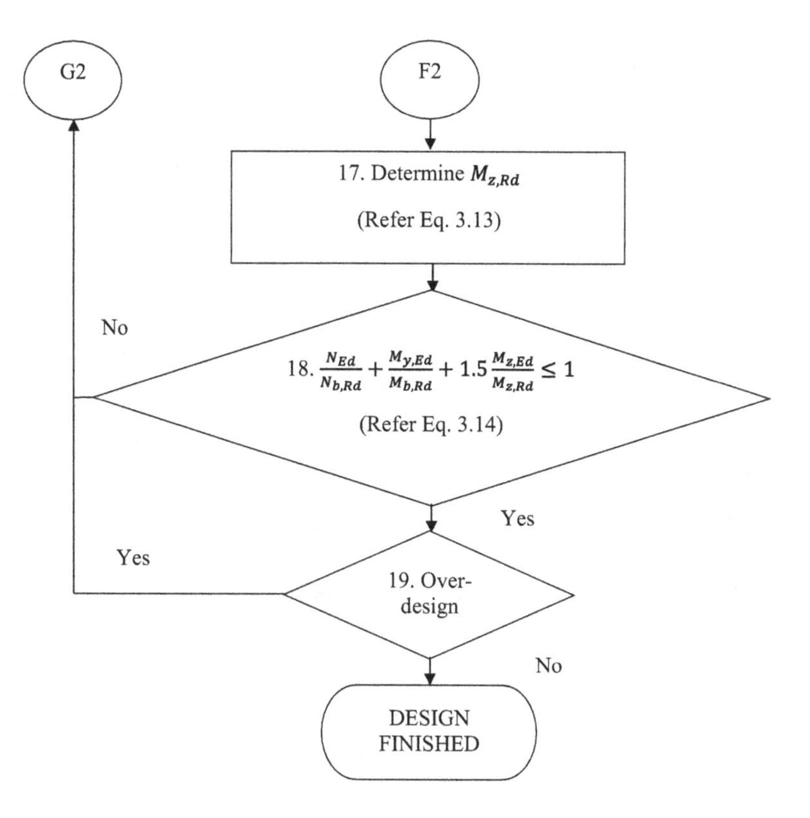

3.2.2 *Example 3-1 Column Design*

Design the 2 m-high column in Fig. 3.3 using the simplified approach. The connection between the column and the beams is pinned, and the bottom end of the column is rigidly connected. Beams A and B sit on 100 mm bearings at each end. The reactions of beams A and B are 100 and 50 kN respectively, while the ultimate load on the column is 10 kN. Steel grade S275 is used for the column (Fig. 3.4).

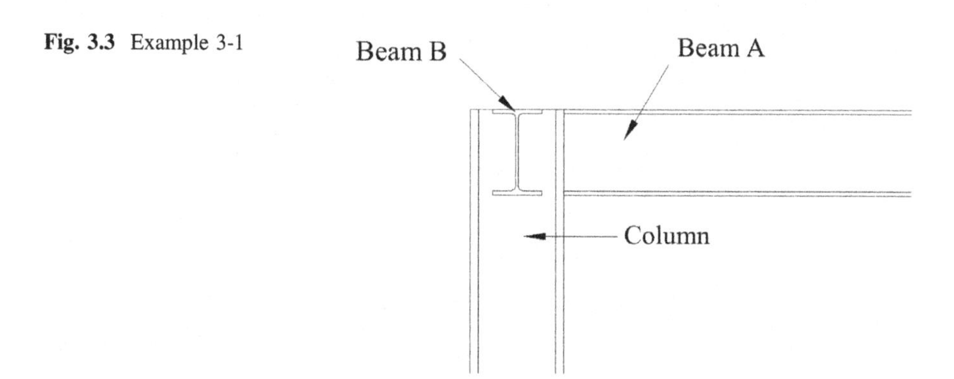

Fig. 3.3 Example 3-1

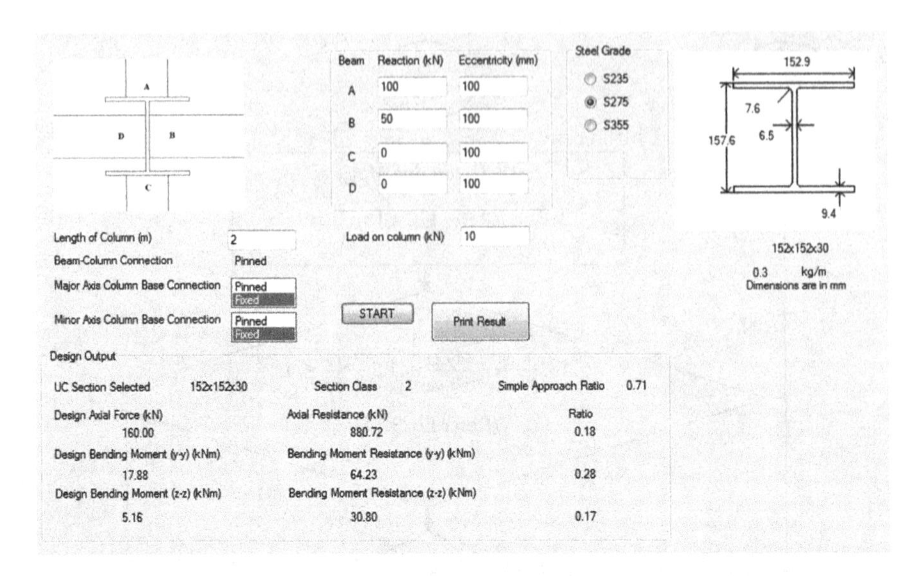

Fig. 3.4 Result for Example 3-1 using steel design based on EC3 program

Step	Reference	Action/calculation	Conclusion
1	References are to BS EN 1993-1-1 unless otherwise stated	Support condition of the column is **fixed-pinned**	
2		Reaction for: beam A = **100 kN** beam B = **50 kN**	$V_{Ed,y\text{-}y}$ = 100 kN $V_{Ed,z\text{-}z}$ =50 kN
3	Table 3.1	Steel grade = **S275** Assume the thicknesses of web and flange are less than 40 mm: **fy = 275 N/mm²**	f_y = 275 N/mm^2
	BS 4 Part 1 2005	Randomly choose a column section for the first trial: Select column section **152 × 152 × 30** The properties of the section is as follows: Depth of section, D = 157.6 mm Width of section, b = 152.9 mm Thickness of web, t_w = 6.5 mm Thickness of flange, t_f = 9.4 mm Root radius, r = 7.6 mm Depth between fillets, d = 123.6 mm Second moment of area about major (y-y) axis, I_y =1748 cm^4 Second moment of area about minor (z-z) axis, I_z =560 cm^4 Radius of gyration about major (y-y) axis, i_y =6.76 cm Radius of gyration about minor (z-z) axis, i_z =3.83 cm Elastic modulus about major (y-y) axis, $W_{el,y}$ =222 cm^3 Elastic modulus about minor (z-z) axis, $W_{el,z}$ =73.3 cm^3	

(continued)

(continued)

Step	Reference	Action/calculation	Conclusion
		Plastic modulus about major (*y-y*) axis, $W_{pl,y}$ =248 cm^3 Plastic modulus about minor (*z-z*) axis, $W_{pl,z}$ =112 cm^3 Warping constant, I_w = 0.031 dm^6 Torsional constant, I_t = 10.5 cm^4 Area of section, A = 38.3 cm^2	
4		$V_{Ed,y\text{-}y}$ = 100 kN $V_{Ed,z\text{-}z}$ = 50 kN Load on column = 10 kN N_{Ed} $=\sum_{i=1}^{n} V_{Ed,i} + load\ on\ column$ =100 + 50 + 10 =**160.00 kN**	N_{Ed} = 160.00 kN
		$M_{y,Ed}$ and $M_{z,Ed}$ can be calculated based on geometry of the column section, as they are induced by eccentricity of loads with respect to centroid of the said section. $M_{y,Ed}$ $= Shear\ difference\ in\ y\text{-}y \times \left(\frac{D}{2} + bearing\ size\right)$ $= 100 \times \left(\frac{157.6\times10^{-3}}{2} + 100 \times 10^{-3}\right)$ = **17.88 kNm**	$M_{y,Ed}$ = 17.88 kNm
		$M_{z,Ed}$ $= Shear\ difference\ in\ z\text{-}z \times \left(\frac{t_w}{2} + bearing\ size\right)$ $= 50 \times \left(\frac{6.5\times10^{-3}}{2} + 100 \times 10^{-3}\right)$ = **5.16 kNm**	$M_{z,Ed}$ = 5.16 kNm
5	Table 5.2	Section classification: i. f_y = 275 N/mm² ε = 0.92 **Class 2** ii. Rolled section, outstand flange: $c = \frac{b-t_w-2r}{2}$ $= \frac{152.9-6.5-2(7.6)}{2}$ = 65.60 mm t_f = 9.4 mm $\frac{c}{t_f} = \frac{65.60}{9.4} = 6.98 < 9\epsilon (=8.28)$ **Class 1** iii. Rolled section, web subjected to compression: $c^* = d$ = 123.6 mm t_w = 6.5 mm $\frac{c*}{t_w} = \frac{123.6}{5.8} = 19.02 < 33\epsilon (=30.36)$ **Class 1** Therefore, the section is **class 2**	Section class 2
6	6.3.1.3(1)	Non-dimensional slenderness can be determined using equation below: $\bar{\lambda} = \frac{L}{i} \times \frac{1}{\pi}\left(\sqrt{\frac{f_y}{E}}\right)$ $= \frac{0.85\times2}{3.83\times10^{-2}} \times \frac{1}{\pi}\left(\sqrt{\frac{275\times10^6}{210\times10^9}}\right)$ = **0.51**	$\bar{\lambda} = 0.51$
7	Table 6.1 Table 6.2	$\frac{h}{b} = \frac{D}{b} = \frac{157.6}{152.9} = 1.03$ t_f = 9.4 mm Determine imperfection factor by consider the following limits: $\frac{h}{b} < 1.2$, $t_f <$ 100 mm and buckling occurs about minor (*z-z*) axis: $\alpha = 0.49$ $\phi = 0.5\left[1 + \alpha(\bar{\lambda} - 0.2) + \bar{\lambda}^2\right]$ $=0.5\left[1 + 0.49 \times (0.51 - 0.2) + (0.51)^2\right]$ =**0.71**	$\phi = 0.71$

(continued)

(continued)

Step	Reference	Action/calculation	Conclusion
8	6.3.1.2(1)	Reduction factor can be determined using equation below: $\chi = \frac{1}{\phi + \sqrt{\phi^2 - \lambda^2}}$ $= \frac{1}{0.71 + \sqrt{(0.71)^2 - (0.51)^2}}$ $= \mathbf{0.83}$	$\chi = 0.83$
9	6.3.1.1(3)	For Class 2 section, $N_{b,Rd} = \frac{\chi A f_y}{\gamma_{M1}}$ $= \frac{0.83 \times 38.3 \times 10^{-4} \times 275 \times 10^6}{1.0}$ $= \mathbf{874.20\ kN}$	$N_{b,Rd}$ = 874.20 kN
10		$\frac{N_{Ed}}{N_{b,Rd}} = \frac{160.00}{874.20} = \mathbf{0.18} < 1$ The buckling resistance of the section is adequate	$\frac{N_{Ed}}{N_{b,Rd}} = 0.18$
11	SN003b Access Steel Document	Critical buckling resistance can be determined using equation below. For pinned-fixed support condition, effective length factor, *K* is taken as 0.85: $M_{cr} = \frac{\pi^2 EI_z}{(KL)^2} \sqrt{\left(\frac{I_w}{I_z} + \frac{(KL)^2 GI_t}{\pi^2 EI_z}\right)}$ $= \frac{\pi^2 \times 210 \times 10^9 \times 560 \times 10^{-8}}{(0.85 \times 2)^2}$ $\times \sqrt{\left(\frac{0.031 \times 10^{-6}}{560 \times 10^{-8}} + \frac{(0.85 \times 2)^2 \times 81 \times 10^9 \times 10.5 \times 10^{-8}}{\pi^2 \times 210 \times 10^9 \times 560 \times 10^{-8}}\right)}$ $= \mathbf{351.35\ kNm}$	M_{cr} = 351.35 kNm
12	6.3.2.2(1)	For Class 2 section, slenderness for lateral torsional buckling can be determined using equation below: $\bar{\lambda}_{LT} = \sqrt{\frac{W_{pl,y} f_y}{M_{cr}}}$ $= \sqrt{\frac{248 \times 10^{-6} \times 275 \times 10^6}{351.35 \times 10^3}}$ $= \mathbf{0.44}$	$\bar{\lambda}_{LT} = 0.44$
13	Table 6.3 Table 6.4	$\frac{h}{b} = \frac{D}{b} = \frac{157.6}{152.9} = 1.03$ Determine imperfection factor: $\frac{h}{b} = 1.03 < 2$ $\alpha_{LT} = 0.21$ $\phi_{LT} = 0.5\left[1 + \alpha_{LT}\left(\bar{\lambda}_{LT} - 0.2\right) + \bar{\lambda}_{LT}^2\right]$ $= 0.5\left[1 + 0.21 \times (0.44 - 0.2) + (0.44)^2\right]$ $= \mathbf{0.62}$	$\phi_{LT} = 0.62$
14	6.3.2.2(1)	Lateral torsional buckling reduction factor can be determined using equation below: $\chi_{LT} = \frac{1}{\phi_{LT} + \sqrt{\phi_{LT}^2 - \bar{\lambda}_{LT}^2}}$ $= \frac{1}{0.62 + \sqrt{(0.62)^2 - (0.44)^2}}$ $= \mathbf{0.95}$	$\chi_{LT} = 0.95$
15	6.3.2.1(3)	For Class 2 section, $M_{b,Rd} = \chi_{LT} W_{pl,y} \frac{f_y}{\gamma_{M1}}$ $= \frac{0.95 \times 248 \times 10^{-6} \times 275 \times 10^6}{1.0}$ $= \mathbf{64.79\ kNm}$	$M_{b,Rd}$ = 64.79 kNm
16		$\frac{M_{y,Ed}}{M_{b,Rd}} = \frac{17.88}{64.79} = \mathbf{0.28} < 1$ The bending resistance of the section is adequate	$\frac{M_{y,Ed}}{M_{b,Rd}} = 0.28$
17	6.2.5(2)	For Class 2 section, $M_{z,Rd} = \frac{W_{pl,z} f_y}{\gamma_{M1}}$ $= \frac{112 \times 10^{-6} \times 275 \times 10^6}{1.0}$ $= \mathbf{30.80\ kNm}$	$M_{z,Rd}$ = 30.80 kNm
18	SN048b-EN-GB Access Steel Document	Check ratio $\frac{N_{Ed}}{N_{b,Rd}} + \frac{M_{y,Ed}}{M_{b,Rd}} + 1.5\frac{M_{z,Ed}}{M_{z,Rd}}$ $= 0.18 + 0.28 + 1.5\left(\frac{5.16}{30.80}\right)$ $= \mathbf{0.71} \leq \mathbf{1}$	$\frac{N_{Ed}}{N_{b,Rd}} + \frac{M_{y,Ed}}{M_{b,Rd}} + 1.5\frac{M_{z,Ed}}{M_{z,Rd}} = 0.71$
19		The ratio is 0.71, which is less than 1. Therefore, the column section 152 × 152 × 30 is **adequate**	

3.2.3 *Example 3-2 Column Design*

Check the suitability of a 254 × 254 × 107 section for the column in Fig. 3.5. Use steel grade S235. The connection between the column and beam is pinned, and the support condition for the base of the column is pinned and fixed about the *y-y* and *z-z* axes respectively (Fig. 3.6).

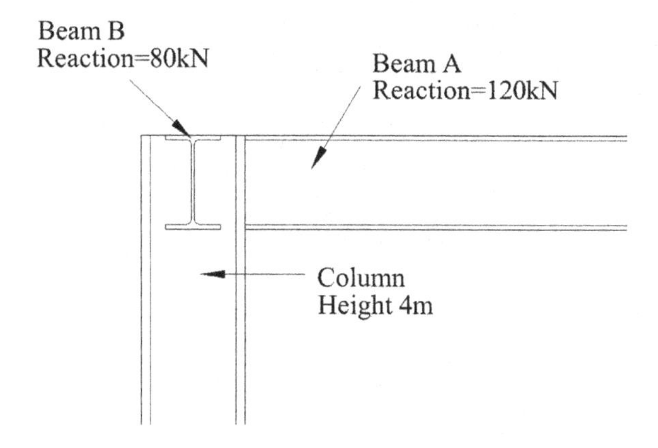

Fig. 3.5 Example 3-2

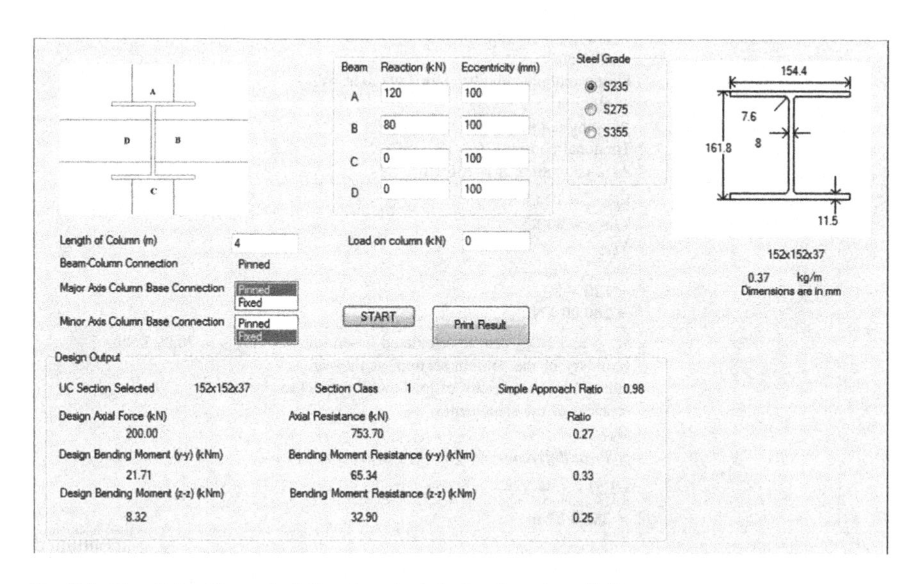

Fig. 3.6 Result for Example 3-2 using steel design based on EC3 program

Step	Reference	Action/calculation	Conclusion
1	References are to BS EN 1993-1-1 unless otherwise stated	Support condition of the column is **pinned-pinned** about *y-y* axis and **fixed-pinned** about *z-z* axis	
2		Reaction for: beam A = **120 kN** beam B = **80 kN**	$V_{Ed,y\text{-}y}$ = 120 kN $V_{Ed,z\text{-}z}$ = 80 kN
3	Table 3.1	Steel grade = **S235** Assume the thicknesses of web and flange are less than 40 mm: $\boldsymbol{fy}$ **= 235 N/mm²**	f_y = 235 N/mm²
	BS 4 Part 1 2005	Try the following column section: Select column section **254** × **254** × **107** The properties of the section is as follows: Depth of section, D = 266.7 mm Width of section, b = 258.8 mm Thickness of web, t_w = 12.8 mm Thickness of flange, t_f = 20.5 mm Root radius, r = 12.7 mm Depth between fillets, d = 200.3 mm Second moment of area about major (*y-y*) axis, I_y =17510 cm⁴ Second moment of area about minor (*z-z*) axis, I_z =5928 cm⁴ Radius of gyration about major (*y-y*) axis, i_y =11.3 cm Radius of gyration about minor (*z-z*) axis, i_z =6.59 cm Elastic modulus about major (*y-y*) axis, $W_{el,y}$ =1313 cm³ Elastic modulus about minor (*z-z*) axis, $\mathrm{W_{el,z}}$ =458 cm³ Plastic modulus about major (*y-y*) axis, $W_{pl,y}$ =1484 cm³ Plastic modulus about minor (*z-z*) axis, $\mathrm{W_{pl,z}}$ = 697 cm³ Warping constant, I_w = 0.898 dm⁶ Torsional constant, I_t = 172 cm⁴ Area of section, A = 136 cm²	
4		$V_{Ed,y\text{-}y}$ = 120 kN $V_{Ed,z\text{-}z}$ = 80 kN N_{Ed} $= \sum_{i=1}^{n} V_{Ed,i}$ = 120 + 80 = **200.00 kN**	N_{Ed} = 200.00 kN
		$M_{y,Ed}$ and $M_{z,Ed}$ can be calculated based on geometry of the column section, as they are induced by eccentricity of loads with respect to centroid of the said section $M_{y,Ed}$ $= Shear\ difference\ in\ y\text{-}y \times \left(\frac{D}{2} + bearing\ size\right)$ $=120 \times \left(\frac{266.7\times10^{-3}}{2} + 100\times10^{-3}\right)$ = **28.00 kNm**	$M_{y,Ed}$ = 28.00 kNm

(continued)

(continued)

		$M_{z,Ed}$ $= Shear\ difference\ in\ z\text{-}z \times \left(\frac{t_w}{2} + bearing\ size\right)$ $= 80 \times \left(\frac{12.8\times10^{-3}}{2} + 100 \times 10^{-3}\right)$ $=$ **8.51 kNm**	$M_{z,Ed}$ = 8.51 kNm
5	Table 5.2	Section classification: i. f_y = 235 N/mm^2 ε = 1 **Class 1** ii. Rolled section, outstand flange: $c = \frac{b-t_w-2r}{2}$ $=\frac{258.8-12.8-2(12.7)}{2}$ =110.3 mm t_f = 20.5 mm $\frac{c}{t_f} = \frac{110.3}{20.5} = 5.38 < 9\epsilon(=9)$ **Class 1** iii. Rolled section, web subjected to compression: $c^* = d$ =200.3 mm t_w = 12.8 mm $\frac{c*}{t_w} = \frac{200.3}{12.8} = 15.65 < 33\epsilon(=33)$ **Class 1** Therefore, the section is **class 1**	Section class 1
6	6.3.1.3(1)	Non-dimensional slenderness can be determined using equation below: $\bar{\lambda} = \frac{KL}{i} \times \frac{1}{\pi}\left(\sqrt{\frac{f_y}{E}}\right)$ Since the support condition for both axes is different, slenderness about each axis should be checked carefully Check slenderness about *y-y* axis $\bar{\lambda} = \frac{1.0\times4}{11.3\times10^{-2}} \times \frac{1}{\pi}\left(\sqrt{\frac{235\times10^6}{210\times10^9}}\right)$ =**0.38** Check slenderness about *z-z* axis $\bar{\lambda} = \frac{0.85\times4}{6.59\times10^{-2}} \times \frac{1}{\pi}\left(\sqrt{\frac{235\times10^6}{210\times10^9}}\right)$ =**0.55** The more critical value should be used, as it governs the resistance of section. Therefore, take slenderness value = **0.55**	$\bar{\lambda} = 0.55$
7	Table 6.1 Table 6.2	$\frac{h}{b} = \frac{D}{b} = \frac{266.7}{258.8} = 1.03$ t_f = 20.5 mm Determine imperfection factor by consider the following limits: $\frac{h}{b} < 1.2$, t_f < 100 mm and buckling occurs about minor (*z-z*) axis: $\alpha = 0.49$ $\phi = 0.5\left[1 + \alpha(\bar{\lambda} - 0.2) + \bar{\lambda}^2\right]$ $=0.5\left[1 + 0.49 \times (0.55 - 0.2) + (0.55)^2\right]$ =**0.74**	$\phi = 0.74$
8	6.3.1.2(1)	Reduction factor can be determined using equation below: $\chi = \frac{1}{\phi + \sqrt{\phi^2 - \bar{\lambda}^2}}$ $= \frac{1}{0.74 + \sqrt{0.74^2 - 0.55^2}}$ =**0.81**	$\chi = 0.81$

(continued)

(continued)

9	6.3.1.1(3)	For Class 1 section, $N_{b,Rd} = \frac{\chi A f_y}{\gamma_{M1}}$ $= \frac{0.81 \times 136 \times 10^{-4} \times 235 \times 10^6}{1.0}$ **=2588.76 kN**	$N_{b,Rd}$ = 2588.76 kN
10		$\frac{N_E}{N_{b,Rd}} = \frac{200.00}{2588.76} = \mathbf{0.08 < 1}$ The buckling resistance of the section is adequate	$\frac{N_{Ed}}{N_{b,Rd}} = 0.18$
11	SN003b Access Steel Document	Critical buckling resistance can be determined using equation below. Since the buckling is occurs about major (*y-y*) axis, support condition about *y-y* axis (pinned-pinned) is considered. In this case, effective length factor, *K* is taken as 1.0: $M_{cr} = \frac{\pi^2 EI_z}{(KL)^2}\sqrt{\left(\frac{I_w}{I_z} + \frac{(KL)^2 GI_t}{\pi^2 EI_z}\right)}$ $= \frac{\pi^2 \times 210 \times 10^9 \times 5928 \times 10^{-8}}{(1.0 \times 4)^2}$ $\times \sqrt{\left(\frac{0.898 \times 10^{-6}}{5928 \times 10^{-8}} + \frac{(1.0 \times 4)^2 \times 81 \times 10^9 \times 172 \times 10^{-8}}{\pi^2 \times 210 \times 10^9 \times 5928 \times 10^{-8}}\right)}$ **=1401.11 kNm**	M_{cr} = 1401.11 kNm
12	6.3.2.2(1)	For Class 1 section, slenderness for lateral torsional buckling can be determined using equation below: $\bar{\lambda}_{LT} = \sqrt{\frac{W_{pl,y} f_y}{M_{cr}}}$ $= \sqrt{\frac{1484 \times 10^{-6} \times 235 \times 10^6}{1401.11 \times 10^3}}$ **=0.50**	$\bar{\lambda}_{LT} = 0.50$
13	Table 6.3 Table 6.4	$\frac{h}{b} = \frac{D}{b} = \frac{266.7}{258.8} = 1.03$ Determine imperfection factor: $\frac{h}{b} = 1.03 < 2$ $\alpha_{LT} = 0.21$ $\phi_{LT} = 0.5\left[1 + \alpha_{LT}\left(\bar{\lambda}_{LT} - 0.2\right) + \bar{\lambda}_{LT}^2\right]$ $=0.5\left[1 + 0.21 \times (0.50 - 0.2) + (0.50)^2\right]$ **=0.66**	$\phi_{LT} = 0.66$
14	6.3.2.2(1)	Lateral torsional buckling reduction factor can be determined using equation below: $\chi_{LT} = \frac{1}{\phi_{LT} + \sqrt{\phi_{LT}^2 - \bar{\lambda}_{LT}^2}}$ $= \frac{1}{0.66 + \sqrt{(0.66)^2 - (0.50)^2}}$ **=0.92**	$\chi_{LT} = 0.92$
15	6.3.2.1(3)	For Class 1 section, $M_{b,Rd} = \chi_{LT} W_{pl,y} \frac{f_y}{\gamma_{M1}}$ $= \frac{0.92 \times 1484 \times 10^{-6} \times 235 \times 10^6}{1.0}$ **=320.84 kNm**	$M_{b,Rd}$ = 320.84 kNm
16		$\frac{M_{y,Ed}}{M_{b,Rd}} = \frac{28.00}{320.84} = 0.09 < 1$ The bending resistance of the section is adequate	$\frac{M_{y,Ed}}{M_{b,Rd}} = 0.09$
17	6.2.5(2)	For Class 1 section, $M_{z,Rd} = \frac{W_{pl,z} f_y}{\gamma_{M1}}$ $= \frac{697 \times 10^{-6} \times 235 \times 10^6}{1.0}$ **=163.80 kNm**	$M_{z,Rd}$ = 163.80 kNm

(continued)

(continued)

18	SN048b-EN-GB Access Steel Document	Check ratio $\frac{N_{Ed}}{N_{b,Rd}} + \frac{M_{y,Ed}}{M_{b,Rd}} + 1.5\frac{M_{z,Ed}}{M_{z,Rd}}$ $=0.07 + 0.09 + 1.5\left(\frac{8.51}{163.80}\right)$ $=\mathbf{0.24} \leq \mathbf{1}$	$\frac{N_{Ed}}{N_{b,Rd}} + \frac{M_{y,Ed}}{M_{b,Rd}} + 1.5\frac{M_{z,Ed}}{M_{z,Rd}} = 0.24$
19		The ratio is 0.24, which is less than 0.5. Therefore, the column section 254 × 254 × 107 is **adequate but not optimum**	

From the program, the optimum section for beam subjected to condition as specified in Example 3.2 is 152 × 152 × 37. This section is obviously smaller than proposed 254 × 254 × 107 section. Therefore, the proposed section is adequate, but not considered as optimum.

3.2.4 Example 3-3 Column Design

Design the 5 m-high column in Fig. 3.7 using the simplified approach. The connections between the column and the beams and the bottom end of the column are pinned. The ultimate load on the column is 6 kN. Steel grade S275 is used for the column.

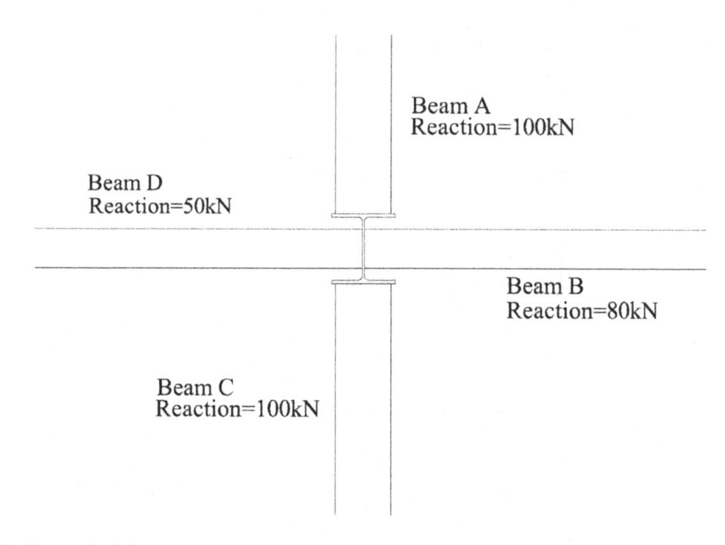

Fig. 3.7 Example 3-3

Step	Reference	Action/calculation	Conclusion
1	References are to BS EN 1993-1-1 unless otherwise stated	Support condition of the column is **pinned-pinned**	
2		Reaction for: beam A = **100 kN** beam B = **80 kN** beam C = **100 kN** beam D = **50 kN** $V_{Ed,y\text{-}y}$ = 100 + 100 = **200 kN** $V_{Ed,z\text{-}z}$ = 80 + 50 = **130 kN**	$V_{Ed,y\text{-}y}$ = 200 kN $V_{Ed,z\text{-}z}$ = 130 kN
3	Table 3.1	Steel grade = **S275** Assume the thicknesses of web and flange are less than 40 mm: ***fy* = 275 N/mm²**	f_y = 275 N/mm²
	BS 4 Part 1 2005	Randomly choose a column section for the first trial: Select column section **203 × 203 × 46** The properties of the section is as follows: Depth of section, D = 203.2 mm Width of section, b = 203.6 mm Thickness of web, t_w = 7.2 mm Thickness of flange, t_f = 11.0 mm Root radius, r = 10.2 mm Depth between fillets, d = 160.8 mm Second moment of area about major (y-y) axis, I_y =4568 cm⁴ Second moment of area about minor (z-z) axis, I_z =1548 cm⁴ Radius of gyration about major (y-y) axis, i_y =8.82 cm Radius of gyration about minor (z-z) axis, i_z =5.13 cm Elastic modulus about major (y-y) axis, $W_{el,y}$ =450 cm³ Elastic modulus about minor (z-z) axis, $W_{el,z}$ =152 cm³ Plastic modulus about major (y-y) axis, $W_{pl,y}$ =497 cm³ Plastic modulus about minor (z-z) axis, $W_{pl,z}$ =231 cm³ Warping constant, I_w = 0.143 dm⁶ Torsional constant, I_t = 22.2 cm⁴ Area of section, A = 58.7 cm²	
4		$V_{Ed,y\text{-}y}$ = 200 kN $V_{Ed,z\text{-}z}$ = 130 kN Load on column = 6 kN $N_{Ed} = \sum_{i=1}^{n} V_{Ed,i} + load\ on\ column$ =200 + 130 + 6 **=336.00 kN**	N_{Ed} = 336.00 kN
		$M_{y,Ed}$ and $M_{z,Ed}$ can be calculated based on geometry of the column section, as they are induced by eccentricity of loads with respect to centroid of the said section $M_{y,Ed} = Shear\ difference\ in\ y\text{-}y \times \left(\frac{D}{2} + bearing\ size\right)$ $=(100 - 100) \times \left(\frac{203.2 \times 10^{-3}}{2} + 100 \times 10^{-3}\right)$ **=0 kNm**	$M_{y,Ed}$ = 0 kNm

(continued)

(continued)

Step	Reference	Action/calculation	Conclusion
		$M_{z,Ed}$ $= \textit{Shear difference in z-z} \times \left(\frac{t_w}{2} + \textit{bearing size}\right)$ $=(80-50)\times\left(\frac{7.2\times10^{-3}}{2}+100\times10^{-3}\right)$ **=3.11 kNm**	$M_{z,Ed}$ = 3.11 kNm
5	Table 5.2	Section classification: i. f_y = 275 N/mm^2 ε = 0.92 **Class 2** ii. Rolled section, outstand flange: $c=\frac{b-t_w-2r}{2}$ $=\frac{203.6-7.2-2(10.2)}{2}$ =88 mm t_f = 11 mm $\frac{c}{t_f}=\frac{88}{11}=8<9\epsilon(=8.28)$ **Class 1** iii. Rolled section, we subjected to compression: $c^*=d$ =160.8 mm t_w = 7.2 mm $\frac{c*}{t_w}=\frac{160.8}{7.2}=22.33<33\epsilon(=30.36)$ **Class 1** Therefore, the section is **class 2**	Section class 2
6	6.3.1.3(1)	Non-dimensional slenderness can be determined using equation below: $\bar{\lambda}=\frac{KL}{i}\times\frac{1}{\pi}\left(\sqrt{\frac{f_y}{E}}\right)$ $=\frac{1.0\times5}{5.13\times10^{-2}}\times\frac{1}{\pi}\left(\sqrt{\frac{275\times10^6}{210\times10^9}}\right)$ **=1.12**	$\bar{\lambda}=1.12$
7	Table 6.1 Table 6.2	$\frac{h}{b}=\frac{D}{b}=\frac{203.2}{203.6}=0.99$ t_f = 11 mm Determine imperfection factor by consider the following limits: $\frac{h}{b}<1.2$, $t_f<100$ mm and buckling occurs about minor (*z-z*) axis: $\alpha=0.49$ $\phi=0.5\left[1+\alpha(\bar{\lambda}-0.2)+\bar{\lambda}^2\right]$ $=0.5\left[1+0.49\times(1.12-0.2)+(1.12)^2\right]$ **=1.35**	$\phi=1.35$
8	6.3.1.2(1)	Reduction factor can be determined using equation below: $\chi=\frac{1}{\phi+\sqrt{\phi^2-\bar{\lambda}^2}}$ $=\frac{1}{1.35+\sqrt{1.35^2-1.12^2}}$ **=0.48**	$\chi=0.48$
9	6.3.1.1(3)	For Class 2 section, $N_{b,Rd}=\frac{\chi A f_y}{\gamma_{M1}}$ $=\frac{0.48\times58.7\times10^{-4}\times275\times10^6}{1.0}$ **=774.84 kN**	$N_{b,Rd}$ = 774.84 kN
10		$\frac{N_{Ed}}{N_{b,Rd}}=\frac{336.00}{774.84}=\mathbf{0.43<1}$ The buckling resistance of the section is adequate	$\frac{N_{Ed}}{N_{b,Rd}}=0.43$

(continued)

(continued)

Step	Reference	Action/calculation	Conclusion
11		*This step is skipped since* $M_{y,Ed}$ *is 0*	
12		*This step is skipped since* $M_{y,Ed}$ *is 0*	
13		*This step is skipped since* $M_{y,Ed}$ *is 0*	
14		*This step is skipped since* $M_{y,Ed}$ *is 0*	
15		*This step is skipped since* $M_{y,Ed}$ *is 0*	
16		*This step is skipped since* $M_{y,Ed}$ *is 0*	
17	6.2.5(2)	For Class 2 section, $M_{z,Rd} = \frac{W_{pl,z}f_y}{\gamma_{M1}}$ $= \frac{231\times10^{-6}\times275\times10^{6}}{1.0}$ **=63.53 kNm**	$M_{z,Rd}$ = 63.53 kNm
18	SN048b-EN-GB Access Steel Document	Check ratio $\frac{N_{Ed}}{N_{b,Rd}} + \frac{M_{y,Ed}}{M_{b,Rd}} + 1.5\frac{M_{z,Ed}}{M_{z,Rd}}$ $=0.43 + 0 + 1.5\left(\frac{3.11}{63.53}\right)$ $\mathbf{=0.50 \leq 1}$	$\frac{N_{Ed}}{N_{b,Rd}} + \frac{M_{y,Ed}}{M_{b,Rd}} + 1.5\frac{M_{z,Ed}}{M_{z,Rd}} = 0.50$
19		The ratio is 0.50, which is less than 1. Therefore, the column section 203 × 203 × 46 is **adequate**, but **barely considered as optimum**	

Step 3 should be repeated and a smaller column section should be chosen for optimum design (Fig. 3.8).

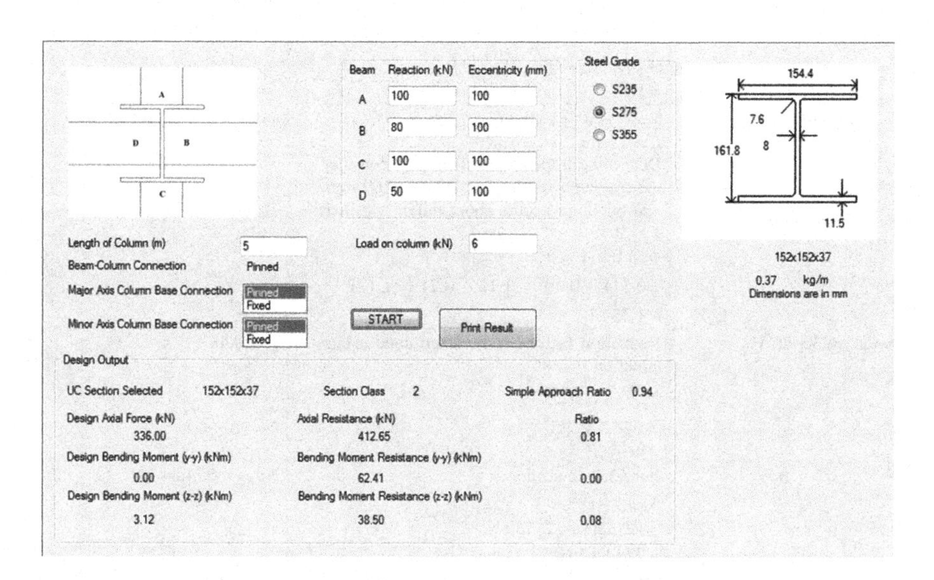

Fig. 3.8 Result for Example 3-3 using steel design based on EC3 program

Step	Reference	Action/calculation	Conclusion
3	Table 3.1	Steel grade = **S275** Assume the thicknesses of web and flange are less than 40 mm: $\boldsymbol{fy = 275}$ **N/mm²**	f_y = 275 N/mm²
	BS 4 Part 1 2005	Select column section **152 × 152 × 37** The properties of the section is as follows: Depth of section, D = 161.8 mm Width of section, b = 154.4 mm Thickness of web, t_w = 8.0 mm Thickness of flange, t_f = 11.5 mm Root radius, r = 7.6 mm Depth between fillets, d = 123.6 mm Second moment of area about major (y-y) axis, I_y =2210 cm⁴ Second moment of area about minor (z-z) axis, I_z =706 cm⁴ Radius of gyration about major (y-y) axis, i_y =6.71 cm Radius of gyration about minor (z-z) axis, i_z =15.5 cm Elastic modulus about major (y-y) axis, $W_{el,y}$ =273 cm³ Elastic modulus about minor (z-z) axis, $W_{el,z}$ =91.5 cm³ Plastic modulus about major (y-y) axis, $W_{pl,y}$ =309 cm³ Plastic modulus about minor (z-z) axis, $W_{pl,z}$ =140 cm³ Warping constant, I_w = 0.04 dm⁶ Torsional constant, I_t = 19.2 cm⁴ Area of section, A = 47.1 cm²	
4		From previous calculation: N_{Ed} = **336.00 kN**	N_{Ed} = 336.00 kN
		$M_{y,Ed}$ and $M_{z,Ed}$ needed to be calculated based on geometry of new column section, as they are induced by eccentricity of loads with respect to centroid of the said section $M_{y,Ed}$ = **0 kNm** since the moment induced by beam A and C cancel out each other	$M_{y,Ed}$ = 0 kNm
		$M_{z,Ed}$ $= \textit{Shear difference in z-z} \times \left(\frac{t_w}{2} + \textit{bearing size}\right)$ $=(80 - 50) \times \left(\frac{8\times10^{-3}}{2} + 100 \times 10^{-3}\right)$ **=3.12 kNm**	$M_{z,Ed}$ = 3.12 kNm
5	Table 5.2	Section classification: i. f_y = 275 N/mm² ε = 0.92 **Class 2** ii. Rolled section, outstand flange: $c = \frac{b-t_w-2r}{2}$ $= \frac{154.4-8-2(7.6)}{2}$ =65.60 mm t_f = 11.5 mm	Section class 2

(continued)

(continued)

Step	Reference	Action/calculation	Conclusion
		$\frac{c}{t_f} = \frac{65.60}{11.5} = 5.70 < 9\epsilon(=8.28)$ **Class 1** iii. Rolled section, web subjected to compression: $c^* = d$ $=123.6$ mm $t_w = 8$ mm $\frac{c*}{t_w} = \frac{123.6}{8} = 15.45 < 33\epsilon(=30.36)$ **Class 1** Therefore, the section is **class 2**	
6	6.3.1.3(1)	Non-dimensional slenderness can be determined using equation below: $\bar{\lambda} = \frac{KL}{i} \times \frac{1}{\pi}\left(\sqrt{\frac{f_y}{E}}\right)$ $= \frac{1.0\times 5}{3.87\times 10^{-2}} \times \frac{1}{\pi}\left(\sqrt{\frac{275\times 10^6}{210\times 10^9}}\right)$ $=$**1.49**	$\bar{\lambda} = 1.49$
7	Table 6.1 Table 6.2	$\frac{h}{b} = \frac{D}{b} = \frac{161.8}{154.4} = 1.05$ $t_f = 11.5$ mm < 100 mm Determine imperfection factor by consider the following limits: $\frac{h}{b} < 1.2$, $t_f <$ 100 mm and buckling occurs about minor (*z-z*) axis: $\alpha = 0.49$ $\phi = 0.5\left[1 + \alpha(\bar{\lambda} - 0.2) + \bar{\lambda}^2\right]$ $=0.5\left[1 + 0.49 \times (1.49 - 0.2) + (1.49)^2\right]$ $=$**1.93**	$\phi = 1.93$
8	6.3.1.2(1)	Reduction factor can be determined using equation below: $\chi = \frac{1}{\phi + \sqrt{\phi^2 - \bar{\lambda}^2}}$ $= \frac{1}{1.93 + \sqrt{1.93^2 - 1.49^2}}$ $=$**0.32**	$\chi = 0.32$
9	6.3.1.1(3)	For Class 2 section, $N_{b,Rd} = \frac{\chi A f_y}{\gamma_{M1}}$ $= \frac{0.32\times 47.1\times 10^{-4}\times 275\times 10^6}{1.0}$ **=414.48 kN**	$N_{b,Rd}$ = 414.48 kN
10		$\frac{N_{Ed}}{N_{b,Rd}} = \frac{336.00}{414.48} = \mathbf{0.81 < 1}$ The buckling resistance of the section is adequate	$\frac{N_{Ed}}{N_{b,Rd}} = 0.81$
11		*This step is skipped since* $M_{y,Ed}$ *is 0*	
12		*This step is skipped since* $M_{y,Ed}$ *is 0*	
13		*This step is skipped since* $M_{y,Ed}$ *is 0*	
14		*This step is skipped since* $M_{y,Ed}$ *is 0*	
15		*This step is skipped since* $M_{y,Ed}$ *is 0*	
16		*This step is skipped since* $M_{y,Ed}$ *is 0*	
17	6.2.5(2)	For Class 2 section, $M_{z,Rd} = \frac{W_{pl,z} f_y}{\gamma_{M1}}$ $= \frac{140\times 10^{-6}\times 275\times 10^6}{1.0}$ **=38.50 kNm**	$M_{z,Rd}$ = 38.50 kNm

(continued)

(continued)

Step	Reference	Action/calculation	Conclusion
18	SN048b-EN-GB Access Steel Document	Check ratio $\frac{N_{Ed}}{N_{b,Rd}}+\frac{M_{y,Ed}}{M_{b,Rd}}+1.5\frac{M_{z,Ed}}{M_{z,Rd}}$ $=0.81+0+1.5\left(\frac{3.12}{38.50}\right)$ $=\mathbf{0.93}\leq\mathbf{1}$	$\frac{N_{Ed}}{N_{b,Rd}}+\frac{M_{y,Ed}}{M_{b,Rd}}+1.5\frac{M_{z,Ed}}{M_{z,Rd}}=0.93$
19		The ratio is 0.93, which is approaching to 1. Therefore, the section 152 × 152 × 37 is **optimum**	

3.3 Exercise: Column Design

3-1 Design the 5 m-high column in Fig. 3.9 using the simplified approach. Use steel grade S235. The connection between the column and the beam is pinned, and the support condition for the base of the column is pinned and fixed about the *y*-*y* and *z*-*z* axes respectively. The ultimate load on the column is 10 kN.

3-2 Design the 5 m-high column in Fig. 3.9 by using the simplified approach. Use steel grade S275. The connections between the column and the beams and the bottom end of the column are pinned. The ultimate load on the column is 10 kN. Compare the result with that obtained in 3-1.

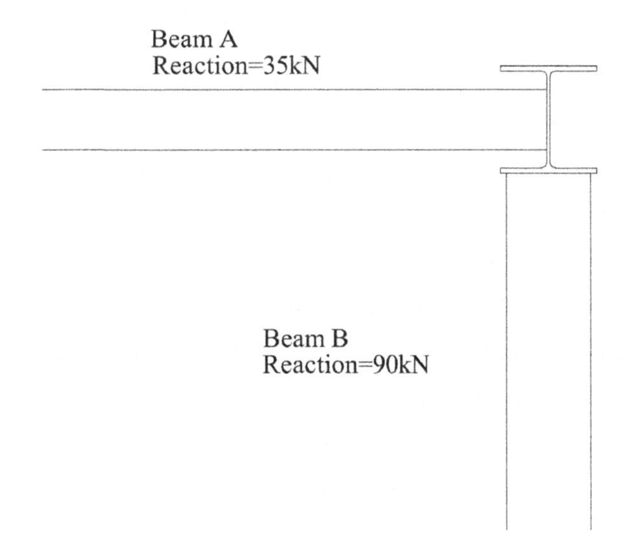

Fig. 3.9 Plan view for Questions 3-1 and 3-2

3-3 Design the 8 m-high column in Fig. 3.10 by using the simplified approach. Use steel grade S275. The connections between the column and the beams and the bottom end of column are pinned.

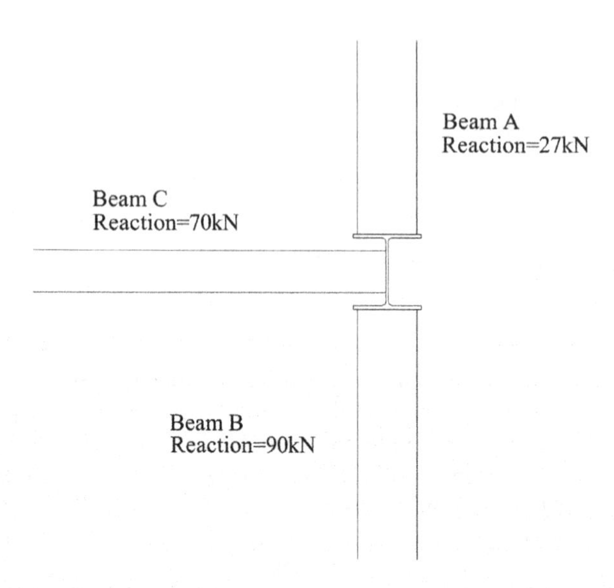

Fig. 3.10 Plan view for Question 3-3

Chapter 4
Connection Design

4.1 Introduction

Connection is a point where two or more different structural members meet. It is important in a frame because it holds all structural members in position and ensures that they behave as a frame. Some examples of connections are beam–beam, beam–column, beam–bracing, and built-up member. Figure 4.1 illustrates some common configurations of steel structure connections.

Connections in steel construction are classified into two common types: welded and bolted.

A welded connection joins two or more structural elements with melted metal. Either arc welding or stick welding may be employed to form a welded connection. Welded connections are generally classified into five types: fillet weld, fillet all-around weld, butt weld, plug weld, and flare groove weld. Figure 4.2 shows the differences among these weld types.

Bolted connection also joins two or more structural elements, but with the use of a fastener, which is secured with the mating of a screw thread, such as in a bolt and nut. Bolted connections have two types: shear connection and tension connection. The type of connection can be determined through the direction of the force acting on the fastener, as shown in Fig. 4.3.

F. Hejazi and T. K. Chun, *Steel Structures Design Based on Eurocode 3*,
https://doi.org/10.1007/978-981-10-8836-0_4

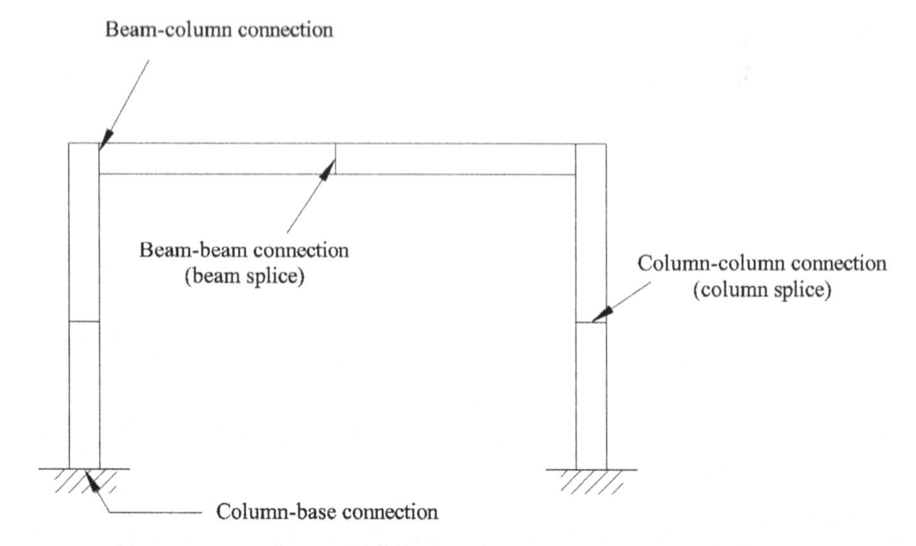

Fig. 4.1 Common configurations of steel structure connection

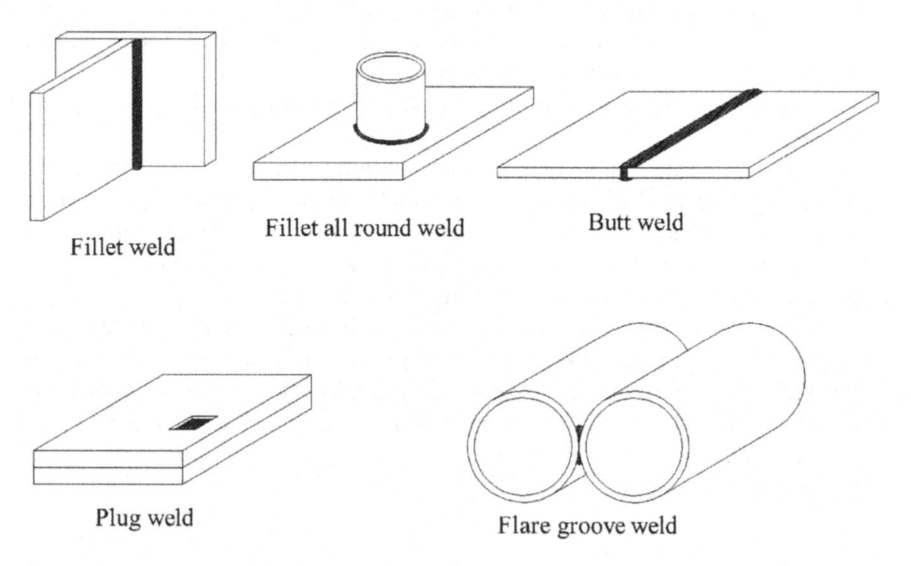

Fig. 4.2 Types of welded connections

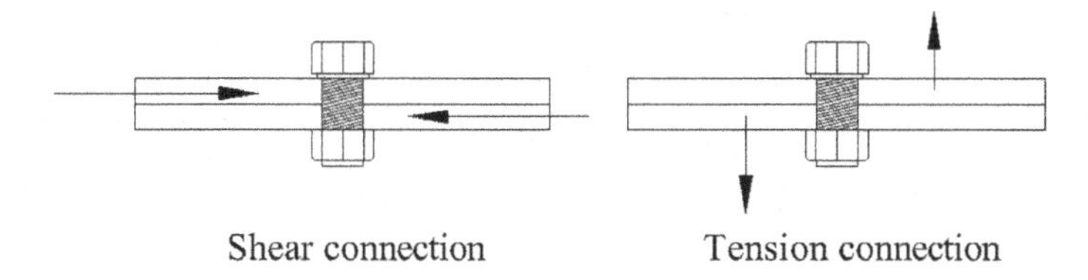

Fig. 4.3 Types of bolted connections

4.2 Design Procedure for a Welded Connection

The design procedure for a welded connection is as follows:

1. Determine the preliminary thickness of the steel welding plate.
2. Select the grade of the plate.
3. Determine the design force N_{Ed} at the joint. If the connection is to be established at the support, then the support reaction should be determined.
4. Determine the preliminary throat thickness a, which is usually defined as $\frac{\sqrt{2}}{2} \times welding\ side$.
5. Determine the correlation factor β_w.
6. Determine the design weld shear strength. The value of γ_{M2} should be set to 1.25.

$$f_{vw,d} = \frac{f_u/\sqrt{3}}{\beta_w \gamma_{M2}} \tag{4.1}$$

where f_u is ultimate tensile strength of steel obtained from Step 2 (Table 4.1)
β_w is correlation factor obtained from Step 5 (Table 4.2)
(BS EN 1993-1-8:2005 4.5.3.3(3))

Table 4.1 Nominal values of yield strength f_y and ultimate tensile strength f_u of hot-rolled structural steel (BS EN 1993-1-1:2005 Table 3.1)

Standard and steel grade (To BS EN 10025-2)	Nominal thickness of element, t (mm)			
	$t \leq 40$ mm		$40\text{ mm} < t \leq 80$ mm	
	f_y (N/mm^2)	f_u (N/mm^2)	f_y (N/mm^2)	f_u (N/mm^2)
S235	235	360	215	360
S275	275	430	255	410
S355	355	490	335	470
S450	440	550	410	550

Table 4.2 Values of the correlation factor β_w for various steel grades (BS EN 1993-1-8:2005 Table 4.1)

Steel grade	β_w
S235	0.8
S275	0.85
S355	0.9
S420	1.0
S460	1.0

7. Determine the weld resistance per length.

$$F_{w,Ed} = f_{vw,d}a \tag{4.2}$$

where $f_{vw,d}$ is design weld shear strength obtained from Step 6 (Eq. 4.1)
a is throat thickness obtained from Step 4

(BS EN 1993-1-8:2005 4.5.3.3(2))

8. Determine the effective welding length by using the equation below. For the edge of a steel plate, the effective welding length is equal to the length of the edge minus $2a$. Specifically, the total welding length should be at least $2a$ more than the computed effective welding length, which depends on the welding pattern. Note that the number of welds manipulates the total welding length. The higher the number of welds, the greater the total welding length.

$$L = \frac{N_{Ed}}{F_{w,Ed}} \tag{4.3}$$

where N_{Ed} is design force at joint obtained from Step 3
$F_{w,Ed}$ is weld resistance per length obtained from Step 7 (Eq. 4.2)

(BS EN 1993-1-8:2005 4.5.3.3(1))

9. Determine the dimension of the steel plate that can provide sufficient welding length. The dimension of the steel plate depends on the number of welds set in Step 8.

4.2.1 Design Flowchart for a Welded Connection

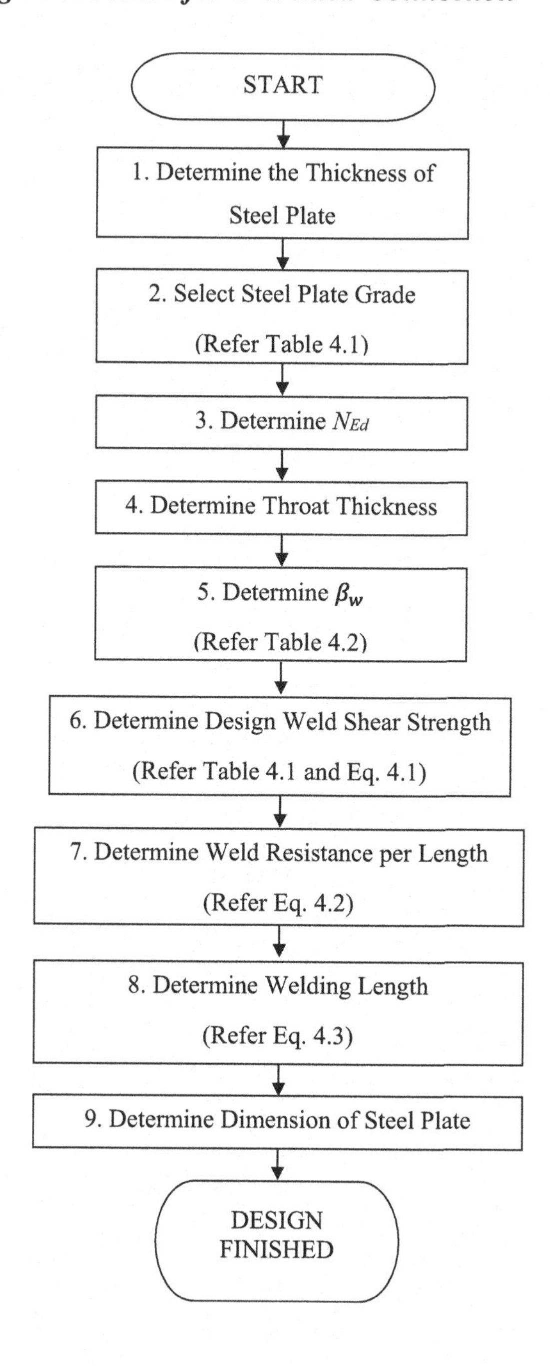

4.2.2 Example 4-1 Welded Connection Design

Find the total welding length of the connection in Fig. 4.4. The load applied to the bracing is 500 kN. Use steel plate grade S235 for the welding plate and the bracing member (Fig. 4.5).

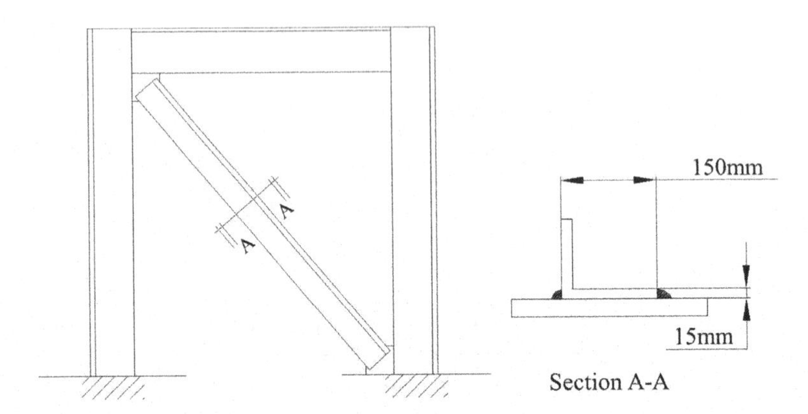

Fig. 4.4 Example 4-1

Step	Reference	Action/calculation	Conclusion
1	References are to BS EN 1993-1-8 unless otherwise stated	From figure above, the thickness of steel bracing member is **15 mm**	t = 15 mm
2	BS EN 1993-1-1 Table 3.1	Steel grade = **S235** t = 15 mm < 40 mm f_u = **360 N/mm²**	f_u = 360 N/mm²
3		N_{Ed} = **500 kN**	N_{Ed} = 500 kN
4		Throat thickness, a $= \frac{\sqrt{2}}{2} \times welding\ side$ $= \frac{\sqrt{2}}{2} t$ $= \frac{\sqrt{2}}{2} \times 15$ = **10.6 mm**	a = 10.6 mm
5	Table 3.1	For steel grade = S235, β_w = **0.8**	β_w = 0.8
6	4.5.3.3(3)	Design weld shear strength. $f_{vw,d}$	$f_{vw,d}$ = 207.8 N/mm²

(continued)

(continued)

Step	Reference	Action/calculation	Conclusion
		$= \frac{f_u/\sqrt{3}}{\beta_w \gamma_{M2}}$ $= \frac{360/\sqrt{3}}{0.8 \times 1.25}$ $= \mathbf{207.8\ N/mm^2}$	
7	4.5.3.3(2)	Weld resistance per length, $F_{w,Ed}$ $= f_{vw,d}a$ $= 207.8 \times 10.6$ $= \mathbf{2.20\ kN/mm}$	$F_{w,Ed}$ = 2.20 kN/mm
8	4.5.3.3(1)	Effective welding length, L $= \frac{N_{Ed}}{F_{wEd}}$ $= \frac{500}{2.20}$ $= \mathbf{227.27\ mm}$	L = 227.27 mm
		From figure below, number of weld is **3** 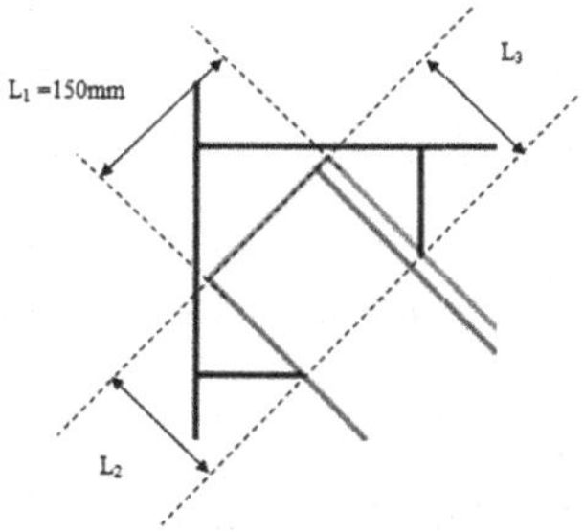 $L_{tot} = L + number\ of\ weld \times 2a$ $= 227.27 + 3 \times 2 \times 10.6$ $= 290.87$ mm $= 291$ mm	L_{tot} = 291 mm
9		From the dimension of bracing member in figure above, L_1 = 150 mm $L_2 = L_3 = \frac{291-150}{2} = 70.5$ mm The minimum welding length at two sides of bracing member is 70.5 mm	

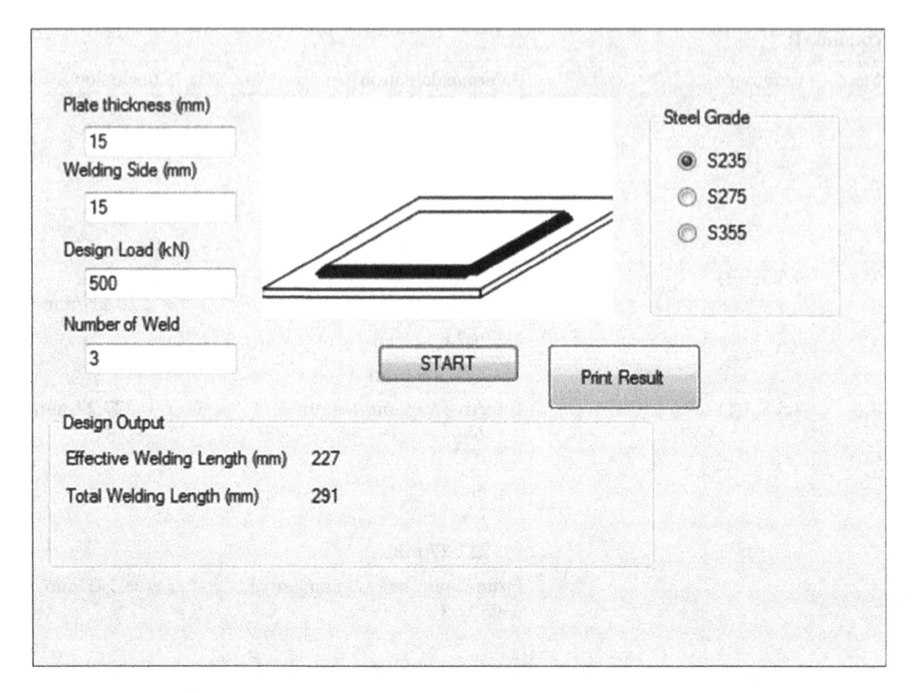

Fig. 4.5 Result for Example 4-1 using steel design based on EC3 program

4.2.3 Example 4-2 Welded Connection Design

Check the suitability of a steel plate for welded connection, which will be established on the left side of the joint (Fig. 4.6). The grade of the steel plate is S235 and the thickness is 10 mm (Fig. 4.7).

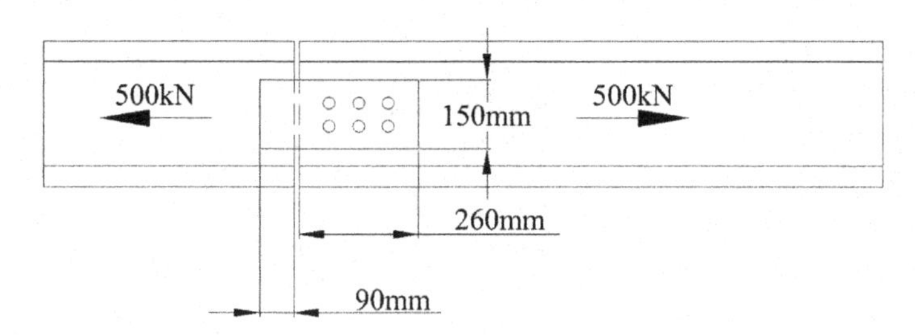

Fig. 4.6 Example 4-2

Step	Reference	Action/calculation	Conclusion
1	References are to BS EN 1993-1-8 unless otherwise stated	The thickness of steel plate is **10 mm**	t = 10 mm
2	BS EN 1993-1-1 Table 3.1	Steel grade = **S235** t = 10 mm < 40 mm f_u = **360 N/mm²**	f_u = 360 N/mm²
3		N_{Ed} = **500 kN**	N_{Ed} = 500 kN
4		Throat thickness, a $= \frac{\sqrt{2}}{2} \times welding\ side$ $= \frac{\sqrt{2}}{2} t$ $= \frac{\sqrt{2}}{2} \times 10$ $= \mathbf{7.1\ mm}$	a = 7.1 mm
5	Table 4.1	For steel grade = S235, β_w = **0.8**	β_w = 0.8
6	4.5.3.3(3)	Design weld shear strength. $f_{vw,d}$ $= \frac{f_u/\sqrt{3}}{\beta_w \gamma_{M2}}$ $= \frac{360/\sqrt{3}}{0.8 \times 1.25}$ $= \mathbf{207.8\ N/mm^2}$	$f_{vw,d}$ = 207.8 N/mm²
7	4.5.3.3(2)	Weld resistance per length, $F_{w,Ed}$ $= f_{vw,d} a$ $= 207.8 \times 7.1$ $= \mathbf{1.48\ kN/mm}$	$F_{w,Ed}$ = 1.48 kN/mm
8	4.5.3.3(1)	Effective welding length, L $= \frac{N_{Ed}}{F_{wEd}}$ $= \frac{500}{1.48}$ $= \mathbf{337.84\ mm}$	L = 337.84 mm
		From figure above, number of weld is **3** $L_{tot} = L + number\ of\ weld \times 2a$ $= 337.84 + 3 \times 2 \times 7.1$ $= 380.44$ mm $= \mathbf{381\ mm}$	L_{tot} = 381 mm
9		From the dimension of welding plate in figure above, the required welding length at two sides of steel plate $= \frac{381 - 150}{2}$ $= 115.5$ mm The minimum welding length at two sides of steel plate is 115.5 mm. However, the available length at two sides of steel plate is only 90 mm. Therefore, the welding plate is **not suitable**	

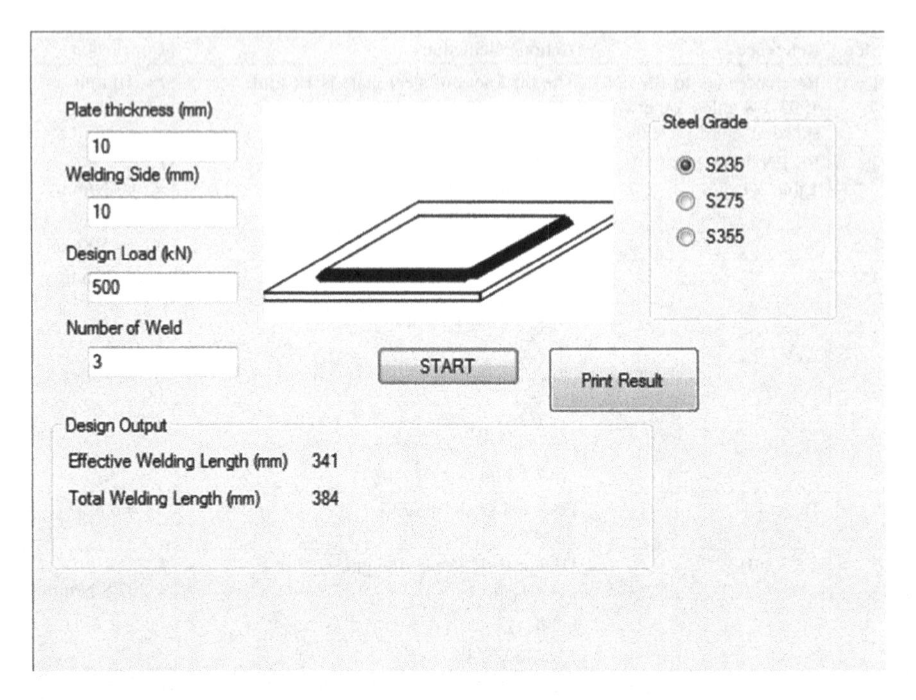

Fig. 4.7 Result for Example 4-2 using steel design based on EC3 program

4.2.4 Example 4-3 Welded Connection Design

Determine the shear resistance of the fillet all-around weld in Fig. 4.8. A steel plate with a grade of S275 and a thickness of 20 mm is used.

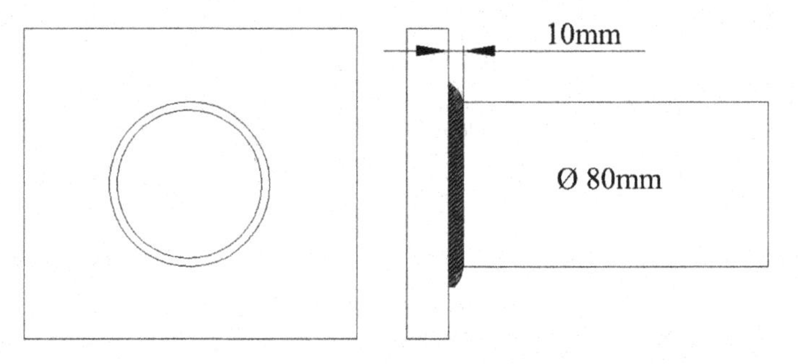

Fig. 4.8 Example 4-3

Step	Reference	Action/calculation	Conclusion
1	References are to BS EN 1993-1-8 unless otherwise stated	Thickness of steel plate is **20 mm** From the figure, welding side is **10 mm**	t = 20 mm welding side = 10 mm

(continued)

(continued)

Step	Reference	Action/calculation	Conclusion
2	BS EN 1993-1-1 Table 3.1	Steel grade = **S275** t = 20 mm < 40 mm f_u = **430 N/mm²**	f_u = 430 N/mm²
3		*This step is skipped as it is not applicable for the situation*	
4		Throat thickness, a $= \frac{\sqrt{2}}{2} \times \textit{welding side}$ $= \frac{\sqrt{2}}{2} t$ $= \frac{\sqrt{2}}{2} \times 10$ $= 7.1$ mm	a = 7.1 mm
5	Table 4.1	For steel grade = S275, β_w = **0.85**	β_w = 0.8
6	4.5.3.3(3)	Design weld shear strength. $f_{vw,d}$ $= \frac{f_u/\sqrt{3}}{\beta_w \gamma_{M2}}$ $= \frac{430/\sqrt{3}}{0.85 \times 1.25}$ $= \mathbf{233.7\ N/mm^2}$	$f_{vw,d}$ = 233.7 N/mm²
7	4.5.3.3(2)	Weld resistance per length, $F_{w,Ed}$ $= f_{vw,d}a$ $= 233.7 \times 7.1$ $= \mathbf{1.65\ kN/mm}$	$F_{w,Ed}$ = 1.65 kN/mm
8	4.5.3.3(1)	Effective welding length, $L = \frac{N_{Ed}}{F_{wEd}}$ Rearrange the equation: Weld resistance, $N_{Ed} = F_{w,Ed} \times L$ Since both ends of the weld is closed, the effective welding length that can be provided is equal to the total welding length $N_{Ed} = 1.65 \times \pi \times 80$ $= 414.69$ kN	N_{Ed} = 414.69 kN
9		*This step is skipped as it is not applicable for the situation*	

4.3 Design Procedure for a Bolted Connection

The design procedure for a bolted connection is as follows:

1. Determine the number of steel plates and their arrangement.
2. Determine the preliminary thickness of the steel plates.
3. Select the grade of the plate (refer to Table 4.1).

Table 4.3 Nominal values of yield strength f_{yb} and ultimate tensile strength f_{ub} of bolts (BS EN 1993-1-8:2005 Table 3.1)

Bolt class	4.6	4.8	5.6	5.8	6.8	8.8	10.9
f_{yb} (N/mm^2)	240	320	300	400	480	640	900
f_{ub} (N/mm^2)	400	400	500	500	600	800	1000

4. Select the bolt class and the bolt diameter. The diameter of a bolt hole d_0 usually equals the bolt diameter plus 2 mm.
5. Determine the design force N_{Ed}. If the connection is to be established at the support, then the support reaction should be determined.
6. Determine the spacing of bolts. The distances between rows of bolts arranged perpendicularly to the direction of the load are denoted by e_1 and P_1, while the distances between rows of bolts arranged parallel to the direction of the load are denoted by e_2 and P_2. The spacing must comply with the limit set in BS EN 1993-1-8. The value of t should be the minimum thickness between the two outermost steel plates.
7. Refer to Table 4.5 to determine the shear resistance per bolt. Next, determine the minimum number of bolts required to resist shear failure by dividing the design force based on the shear resistance per bolt.

Table 4.4 Minimum and maximum spacing, end distances and edge distances (BS EN 1993-1-8:2005 Table 3.3)

Distance and spacing	Minimum	Maximum		
		Structures made from steels conforming to EN10025 except to EN10025-5		Structures made from steel conforming to EN10025-5
		Steel exposed to the weather or other corrosive influences	Steel not exposed to the weather or other corrosive influences	Steel used unprotected
End distance e_1	$1.2d_0$	$4t + 40$ mm		Larger of $8t$ or 125 mm
Edge distance e_2	$1.2d_0$	$4t + 40$ mm		Larger of $8t$ or 125 mm
Spacing p_1	$2.2d_0$	Smaller of $14t$ or 200 mm	Smaller of $14t$ or 200 mm	Smaller of $14t_{min}$ or 175 mm
Spacing $p_{1,0}$		Smaller of $14t$ or 200 mm		
Spacing $p_{1,i}$		Smaller of $28t$ or 200 mm		
Spacing p_2	$2.4d_0$	Smaller of $14t$ or 200 mm	Smaller of $14t$ or 200 mm	Smaller of $14t_{min}$ or 175 mm

Where d_0 is diameter of bolt hole obtained from Step 4
t is minimum thickness between the two outermost steel plates obtained from Step 2

Table 4.5 Design resistance for individual fasteners subjected to shear and/or tension (BS EN 1993-1-8:2005 Table 3.4)

Shear resistance per shear plane	$F_{v,Rd} = \frac{a_v f_{ub} A}{\gamma_{M2}}$ where $a_v = \begin{cases} 0.5, \textit{Bolt class } 4.8, 5.8, 6.8, 10.9 \\ 0.6, \textit{Bolt class } 4.6, 5.6, 8.8 \end{cases}$ A = cross sectional area of bolt
Bearing resistance	$F_{b,Rd} = \frac{k_1 a_b f_u d t}{\gamma_{M2}}$ where (conservatively) $a_b = \min\left\{\frac{e_1}{3d_0}; \frac{P_1}{3d_0} - \frac{1}{4}; \frac{f_{ub}}{f_u}; 1.0\right\}$ $k_1 = \min\left\{2.8\frac{e_2}{d_0} - 1.7; 1.4\frac{P_2}{d_0} - 1.7; 2.5\right\}$
Tension resistance	$F_{t,Rd} = \frac{k_2 f_{ub} A}{\gamma_{M2}}$ where k_2 = 0.9 for normal bolts

Where
f_{ub} is ultimate tensile of bolt obtained from Step 4 (Table 4.3)
d is diameter of bolt obtained from Step 4
d_0 is diameter of bolt hole obtained from Step 4
e_1, p_1, e_2, p_2 are spacing obtained from Step 6 (Table 4.4)

8. Refer to Table 4.5 to determine the bearing resistance per bolt. The value of t should be the minimum between the summations of the steel plate thicknesses in both directions. Next, determine the minimum number of bolts required to resist bearing failure by dividing the design force based on the bearing resistance per bolt.
9. Refer to Table 4.5 to determine the tension resistance per bolt. Next, determine the minimum number of bolts required to resist tensile failure by dividing the design force based on the tension resistance per bolt.
10. Determine the number of bolts required for the situation by selecting the maximum number of bolts required obtained in Steps 7, 8, and 9.
11. Check the ratio of design force to shear resistance, bearing resistance, and tension resistance based on the number of bolts obtained in Step 10.

4.3.1 Design Flowchart for a Bolted Connection

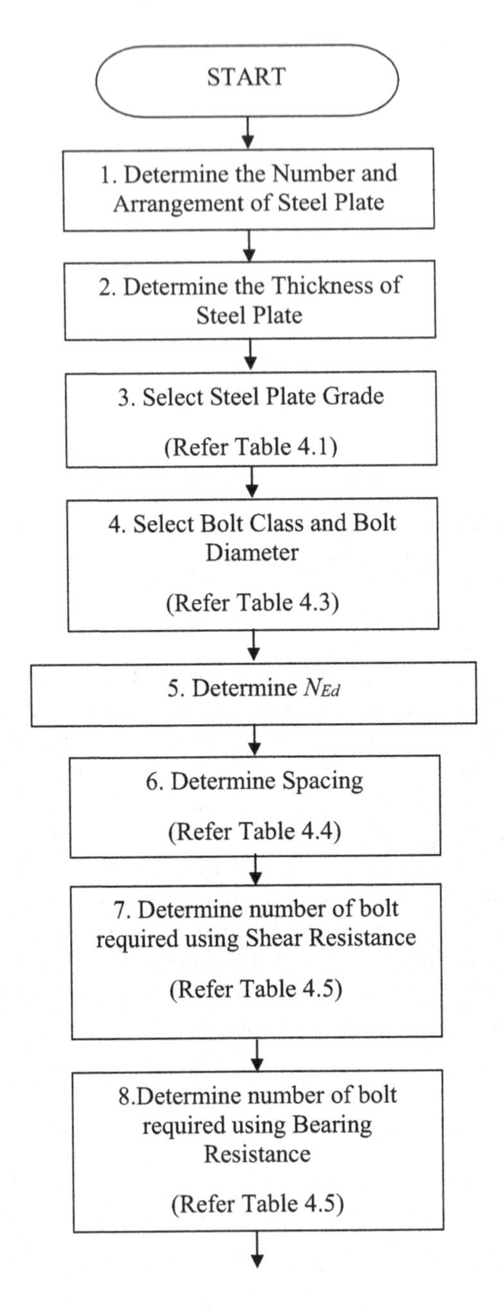

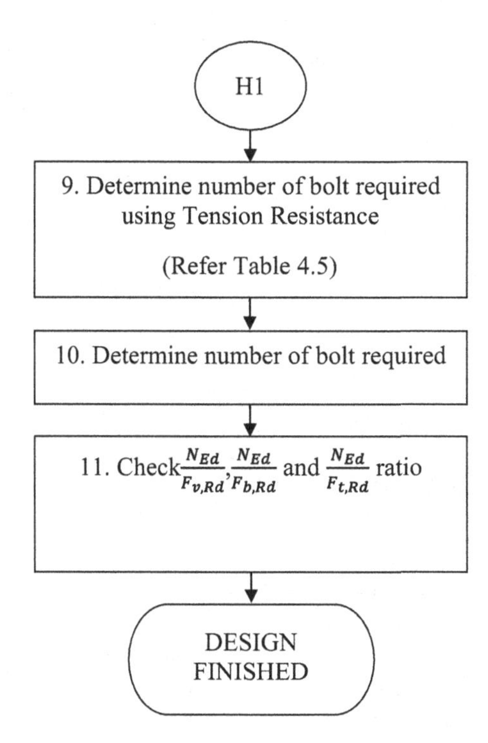

4.3.2 *Example 4-4 Bolted Connection Design*

Check the suitability of the bolt arrangement in Fig. 4.9 if the joint is designed to carry 100 kN. The diameter and the class of bolts are 20 mm and 10.9 respectively. The grade of the steel plate used is S235 (Fig. 4.10).

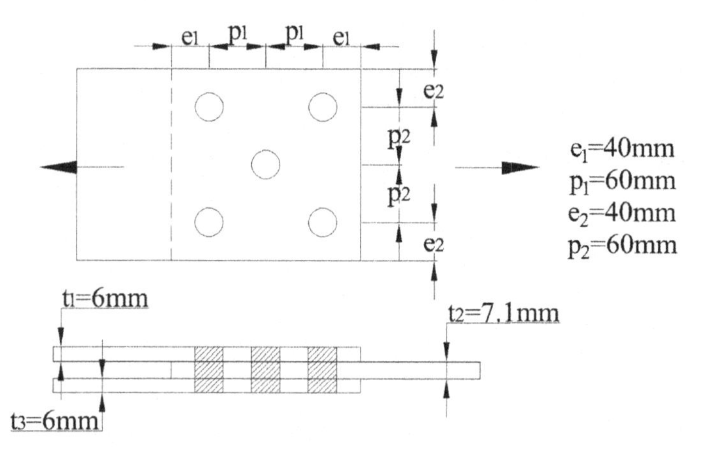

Fig. 4.9 Example 4-4

Step	Reference	Action/calculation	Conclusion
1	References are to BS EN 1993-1-8 unless otherwise stated	Number of plate = **3**, arranged in a way as shown in figure above	Number of plate = 3
2		Thickness of each steel plate is as shown in figure above	$t_1 = 6$ mm $t_2 = 7.1$ mm $t_3 = 6$ mm
3	BS EN 1993-1-1 Table 3.1	Steel grade = **S235** The thicknesses of steel plates are less than 40 mm $\boldsymbol{f_u}$ **= 360 N/mm²**	$f_u = 360$ N/mm²
4	Table 3.1	Bolt class = **10.9**, $\boldsymbol{f_{ub}}$ **= 1000 N/mm²** Bolt diameter, d = **20 mm** Hole diameter, d_0 = 20 + 2 = **22 mm**	Bolt class = 10.9 f_{ub} = 1000 N/mm² d = 20 mm d_0 = 22 mm
5		N_{Ed} = **100 kN**	N_{Ed} = 100 kN
6	Table 3.3	Minimum spacing for: $e_1 = 1.2d_0 = 1.2 \times 22 = \mathbf{26.4\ mm}$ $e_2 = 1.2d_0 = 1.2 \times 22 = \mathbf{26.4\ mm}$ $p_1 = 2.2d_0 = 2.2 \times 22 = \mathbf{48.4\ mm}$ $p_2 = 2.4d_0 = 2.4 \times 22 = \mathbf{52.8\ mm}$ Maximum spacing for: $e_1 = 4t + 40 = 4 \times 6 + 40 = \mathbf{64\ mm}$ $e_2 = 4t + 40 = 4 \times 6 + 40 = \mathbf{64\ mm}$ $p_1 = \min\{14t; 200\} = \min\{14 \times 6; 200\} = \mathbf{84\ mm}$ $p_2 = \min\{14t; 200\} = \min\{14 \times 6; 200\} = \mathbf{84\ mm}$ Compare spacing given with respective upper and lower limit: e_1: 26.4 mm < **40 mm** < 64 mm e_2: 26.4 mm < **40 mm** < 64 mm p_1: 48.4 mm < **60 mm** < 84 mm p_2: 52.8 mm < **60 mm** < 84 mm ∴ *The spacings set are* ***adequate***	e_1 = 40 mm e_2 = 40 mm p_1 = 60 mm p_2 = 60 mm
7	Table 3.4		Number of bolt = 5

(continued)

(continued)

Step	Reference	Action/calculation	Conclusion
		From the figure, the number of bolt provided is **5** Therefore, determine the shear, bearing and tensile resistance of the bolted connection instead	
		For bolt class 10.9, a_v = **0.5** For d = 20 mm, $A = \frac{\pi d^2}{4} = \frac{\pi \times 20^2}{4}$ $= 314.16\ \text{mm}^2$ Number of shear plane = Number of plate – 1 = 3 – 1 = **2** Individual shear resistance per shear plane, $F_{v,Rd}$ $= \frac{a_v f_{ub} A}{\gamma_{M2}}$ $= \frac{0.5 \times 1000 \times 314.16}{1.25}$ = 125.66 kN *Total* $F_{v,Rd}$ = *Individual* $F_{v,Rd}$ × *shear plane* × *bolt number* = 125.66 × 2 × 5 = **1256.6 kN**	$F_{v,Rd}$ = 1256.6 kN
8	Table 3.4	Conservatively, $a_b = \min\left\{\frac{e_1}{3d_0}; \frac{P_1}{3d_0} - \frac{1}{4}; \frac{f_{ub}}{f_u}; 1.0\right\}$ $= \min\left\{\frac{40}{3 \times 22}; \frac{60}{3 \times 22} - \frac{1}{4}; \frac{1000}{360}; 1.0\right\}$ $= \min\{0.61; 0.66; 2.78; 1.0\}$ = **0.61** $k_1 = \min\left\{2.8\frac{e_2}{d_0} - 1.7; 1.4\frac{P_2}{d_0} - 1.7; 2.5\right\}$ $= \min\left\{2.8 \times \frac{40}{22} - 1.7; 1.4 \times \frac{60}{22} - 1.7; 2.5\right\}$ $= \min\{3.39; 2.12; 2.5\}$ = **2.12**	$F_{b,Rd}$ = 264.3 kN

(continued)

(continued)

Step	Reference	Action/calculation	Conclusion
		Individual bearing resistance, $F_{b,Rd}$ $= \frac{k_1 a_b f_u d t}{\gamma_{M2}}$ $= \frac{2.12 \times 0.61 \times 360 \times 20 \times 7.1}{1.25}$ $=$ **52.89 kN** *Total* $F_{b,Rd}$ $=$ *Individual* $F_{b,Rd} \times$ *bolt number* $= 52.89 \times 5$ $=$ **264.3 kN**	
9	Table 3.4	Individual tension resistance, $F_{t,Rd}$ $= \frac{k_2 f_{ub} A}{\gamma_{M2}}$ $= \frac{0.9 \times 1000 \times 314.16}{1.25}$ $=$ **226.20 kN** *Total* $F_{t,Rd}$ $=$ *Individual* $F_{t,Rd} \times$ *bolt number* $= 226.20 \times 5$ $=$ **1131.0 kN**	$F_{t,Rd} = 1131.0$ kN
10		*This step is skipped as it is not applicable for the situation*	
11		Check the following ratio: $\frac{N_{Ed}}{F_{v,Rd}} = \frac{100}{1256.6} = \mathbf{0.08}$ $\frac{N_{Ed}}{F_{b,Rd}} = \frac{100}{264.3} = \mathbf{0.38}$ $\frac{N_{Ed}}{F_{t,Rd}} = \frac{100}{1131.0} = \mathbf{0.09}$ None of these ratios exceed 0.5. This means although the bolt arrangement can support the load, but it is considered **over-design** for this case	

From the program, it is found that with proposed parameters specified in Example 4-4, 2 bolts are sufficient to resist the design load. However, the number of bolt proposed in Example 4-4 is 5. This indicates the proposed bolt arrangement is overdesigned.

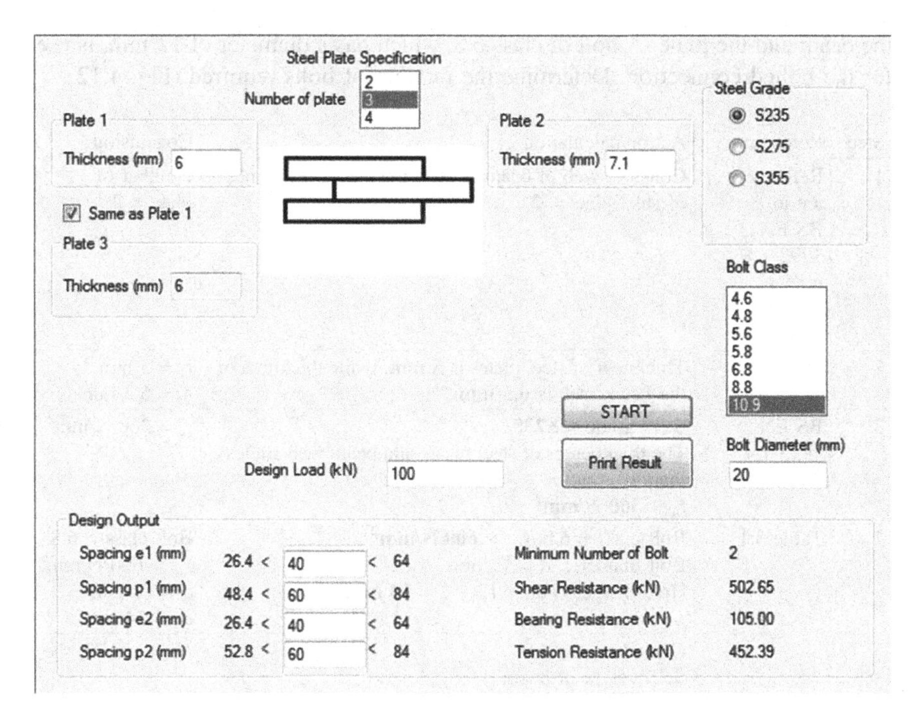

Fig. 4.10 Result for Example 4-4 using steel design based on EC3 program

4.3.3 Example 4-5 Bolted Connection Design

A shear splice is assigned at point B using bolts and a steel plate (Fig. 4.11). The dimension of the beam section is 254 × 146 × 37, and steel grade S235 is used for

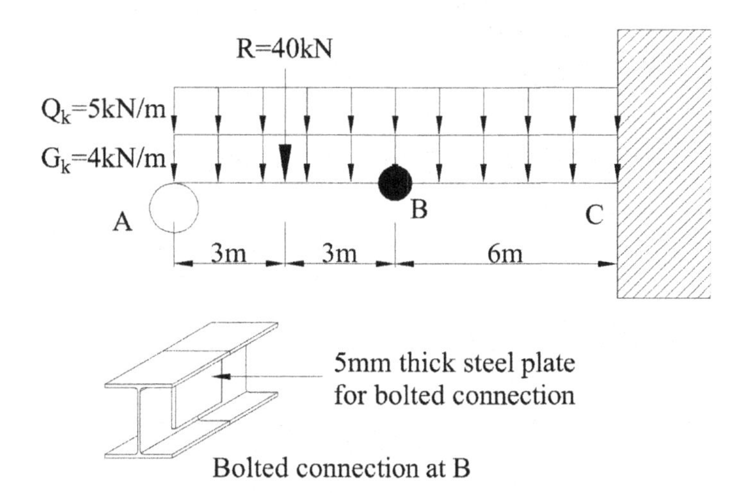

Fig. 4.11 Example 4-5

the beam and the plate. A bolt of class 6.8, which has a diameter of 12 mm, is used for the bolted connection. Determine the number of bolts required (Fig. 4.12).

Step	Reference	Action/calculation	Conclusion
1	References are to BS EN 1993-1-8 unless otherwise stated	Consider web of beam as steel plate as well, number of steel plate = **2**	Number of plate = 2
2		Thickness of steel plates is **5 mm**, while thickness of the beam web is **6.3 mm**	$t_1 = 5$ mm $t_2 = 6.3$ mm
3	BS EN 1993-1-1 Table 3.1	Steel grade = **S235** The thicknesses of steel plates and beam web are less than 40 mm: f_u **= 360 N/mm²**	$f_u = 360\ \text{N/mm}^2$
4	Table 3.1	Bolt class = **6.8**, f_{ub} **= 600 N/mm²** Bolt diameter, d = **12 mm** Hole diameter, $d_0 = 12 + 2 =$ **14 mm**	Bolt class = 6.8 $f_{ub} = 600\ \text{N/mm}^2$ $d = 12$ mm $d_0 = 14$ mm
5		Consider span AB Self-weight of beam = 37 kg/m × 9.81 N/kg = **0.36 kN/m** For ULS, partial factor of safety for both permanent action and variable action selected are 1.35 and 1.5 respectively Uniformly distributed load, w_{ult} $= 1.35G_k + 1.5Q_k$ $= 1.35(5 + 0.36) + 1.5(4)$ = **13.24 kN/m** By principle of superposition, V_{Ed} for simply supported beam (span AB) can be determined using equation below: V_{Ed} (at point B) $= \frac{w_{ult}L}{2} + \frac{R}{2}$ $= \frac{13.24 \times 6}{2} + \frac{40}{2}$ = 59.72 kN	$N_{Ed} = 59.72$ kN

(continued)

(continued)

Step	Reference	Action/calculation	Conclusion
		$N_{Ed} = V_{Ed} = \mathbf{59.72\ kN}$	
6	Table 3.3	Minimum spacing for: $e_1 = 1.2d_0 = 1.2 \times 14 = \mathbf{16.8\ mm}$ $e_2 = 1.2d_0 = 1.2 \times 14 = \mathbf{16.8\ mm}$ $p_1 = 2.2d_0 = 2.2 \times 14 = \mathbf{30.8\ mm}$ $p_2 = 2.4d_0 = 2.4 \times 14 = \mathbf{33.6\ mm}$ Maximum spacing for: $e_1 = 4t + 40 = 4 \times 5 + 40 = \mathbf{60\ mm}$ $e_2 = 4t + 40 = 4 \times 5 + 40 = \mathbf{60\ mm}$ $p_1 = \min\{14t; 200\} = \min\{14 \times 5; 200\} = \mathbf{70\ mm}$ $p_2 = \min\{14t; 200\} = \min\{14 \times 5; 200\} = \mathbf{70\ mm}$ Try following spacing: $e_1 = \mathbf{20\ mm}$ $e_2 = \mathbf{20\ mm}$ $p_1 = \mathbf{40\ mm}$ $p_2 = \mathbf{40\ mm}$ The depth between fillet for 254 × 146 × 37 beam section is 216 mm, while the vertical dimension of proposed steel plate for bolted connection is 2 $(e_2 + p_2)$, which is 160 mm and it can fit between the fillet	$e_1 = 20$ mm $e_2 = 20$ mm $p_1 = 40$ mm $p_2 = 40$ mm
7	Table 3.4	For bolt class 6.8, $a_v = \mathbf{0.5}$ For $d = 12$ mm, $A = \frac{\pi d^2}{4} = \frac{\pi \times 12^2}{4}$ $= \mathbf{113.10\ mm^2}$ Number of shear plane = Number of plate − 1 = 2 − 1 = **1** Individual shear resistance per shear plane, $F_{v,Rd}$ $= \frac{a_v f_{ub} A}{\gamma_{M2}}$ $= \frac{0.5 \times 600 \times 113.10}{1.25}$ $= \mathbf{27.14\ kN}$ *$F_{v,Rd}$ per bolt* *= Individual $F_{v,Rd}$ × shear plane* = 27.14 × 1 = **27.14** kN	Number of bolt for shear resistance = 3

(continued)

(continued)

Step	Reference	Action/calculation	Conclusion
		Number of bolt required $= \frac{N_{Ed}}{F_{v,Rd}} = \frac{59.72}{27.14} = 2.2 = \mathbf{3}$	
8	Table 3.4	Conservatively, $a_b = \min\left\{\frac{e_1}{3d_0}; \frac{P_1}{3d_0} - \frac{1}{4}; \frac{f_{ub}}{f_u}; 1.0\right\}$ $= \min\left\{\frac{20}{3 \times 14}; \frac{40}{3 \times 14} - \frac{1}{4}; \frac{600}{360}; 1.0\right\}$ $= \min\{0.48; 0.70; 1.67; 1.0\}$ $= \mathbf{0.48}$ $k_1 = \min\left\{2.8\frac{e_2}{d_0} - 1.7; 1.4\frac{P_2}{d_0} - 1.7; 2.5\right\}$ $= \min\left\{2.8 \times \frac{20}{14} - 1.7; 1.4 \times \frac{40}{14} - 1.7; 2.5\right\}$ $= \min\{2.3; 2.3; 2.5\}$ $= \mathbf{2.3}$ Individual bearing resistance, $F_{b,Rd} = \frac{k_1 a_b f_u dt}{\gamma_{M2}} = \frac{2.3 \times 0.48 \times 360 \times 12 \times 5}{1.25}$ $= \mathbf{19.08\ kN}$ *Number of bolt required* $= \frac{N_{Ed}}{F_{b,Rd}} = \frac{59.72}{19.08} = 3.1 = \mathbf{4}$	Number of bolt for bearing resistance = 4
9	Table 3.4	Individual tension resistance, $F_{t,Rd} = \frac{k_2 f_{ub} A}{\gamma_{M2}} = \frac{0.9 \times 600 \times 113.10}{1.25}$ $= \mathbf{48.86\ kN}$	Number of bolt for tensile resistance = 2

(continued)

(continued)

Step	Reference	Action/calculation	Conclusion
		Number of bolt required $= \frac{N_{Ed}}{F_{t,Rd}}$ $= \frac{59.72}{48.86}$ $= 1.2 = \mathbf{2}$	
10		Number of bolt required = **4**	Number of bolt = 4
11		Check the following ratio: $\frac{N_{Ed}}{F_{v,Rd}} = \frac{59.72}{27.14 \times 4} = 0.55$ $\frac{N_{Ed}}{F_{b,Rd}} = \frac{59.72}{19.08 \times 4} = 0.78$ $\frac{N_{Ed}}{F_{t,Rd}} = \frac{59.72}{48.86 \times 4} = 0.31$	

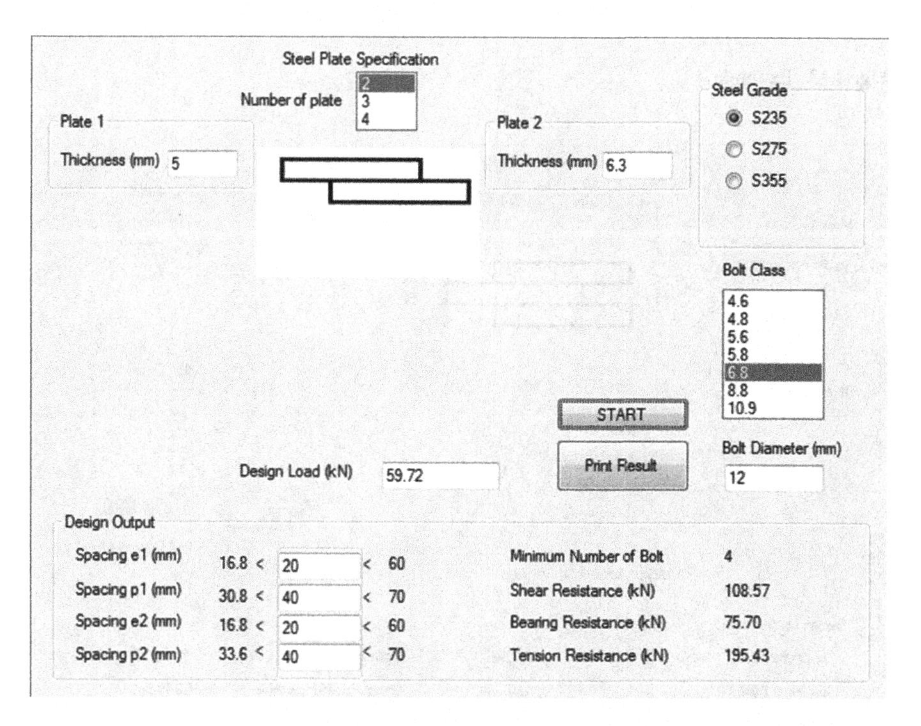

Fig. 4.12 Result for Example 4-5 using steel design based on EC3 program

4.3.4 Example 4-6 Bolted Connection Design

Check the suitability of a 200 mm × 500 mm × 7 mm steel plate in establishing a bolted connection at a beam splice (Fig. 4.13). The steel grade is S235, the bolt class is 10.9, and the bolt diameter is 24 mm. The beam web is 18.4 mm thick Fig. 4.14.

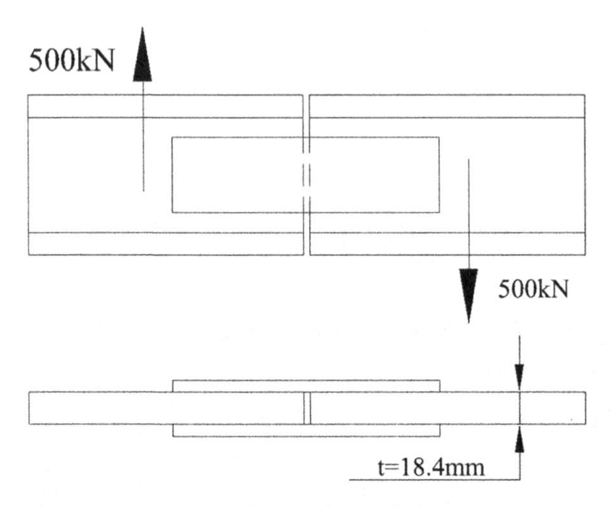

Fig. 4.13 Example 4-6

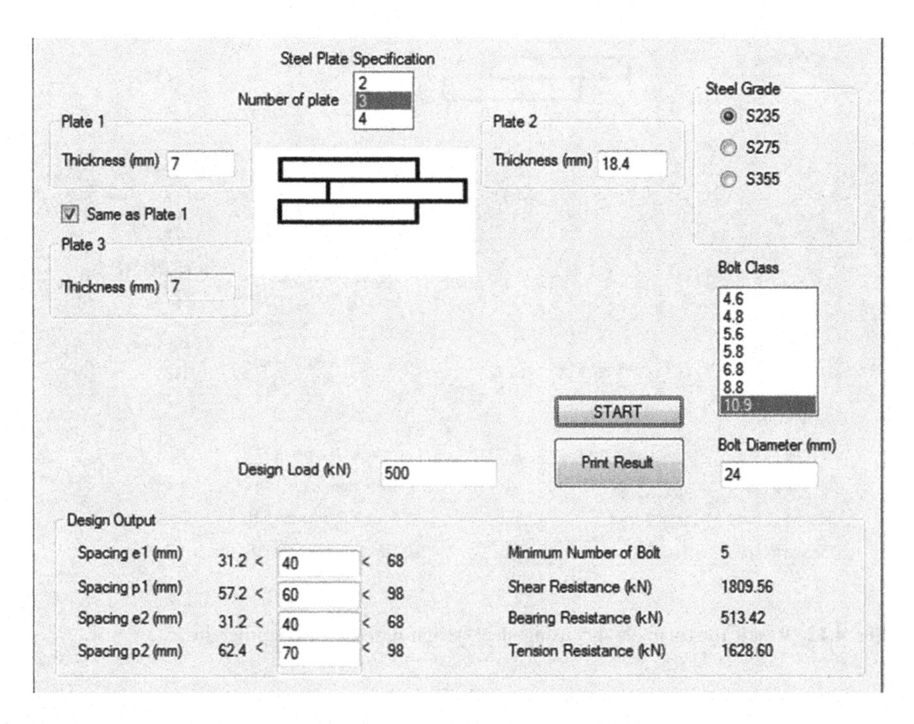

Fig. 4.14 Result for Example 4-6 using steel design based on EC3 program

Step	Reference	Action/calculation	Conclusion
1	References are to BS EN 1993-1-8 unless otherwise stated	Consider web of beam as steel plate as well, number of steel plate = **3**	Number of plate = 3
2		Thickness of steel plates is **7 mm**, while thickness of the beam web is **18.4 mm**	t_1 = 7 mm t_2 = 18.4 mm t_3 = 7 mm
3	BS EN 1993-1-1 Table 3.1	Steel grade = **S235** The thicknesses of steel plates and web are less than 40 mm: f_u = **360 N/mm²**	f_u = 360 N/mm²
4	Table 3.1	Bolt class = **10.9**, f_{ub} = **1000 N/mm²** Bolt diameter, d = **24 mm** Hole diameter, d_0 = 24 + 2 = **26 mm**	Bolt class = 10.6 f_{ub} = 1000 N/mm² d = 24 mm d_0 = 26 mm
5		N_{Ed} = **500 kN**	N_{Ed} = 500 kN
6	Table 3.3	Minimum spacing for: $e_1 = 1.2d_0 = 1.2 \times 26 = \mathbf{31.2\ mm}$ $e_2 = 1.2d_0 = 1.2 \times 26 = \mathbf{31.2\ mm}$ $p_1 = 2.2d_0 = 2.2 \times 26 = \mathbf{57.2\ mm}$ $p_2 = 2.4d_0 = 2.4 \times 26 = \mathbf{62.4\ mm}$ Maximum spacing for: $e_1 = 4t + 40 = 4 \times 7 + 40 = \mathbf{68\ mm}$ $e_2 = 4t + 40 = 4 \times 7 + 40 = \mathbf{68\ mm}$ $p_1 = \min\{14t; 200\} = \min\{14 \times 7; 200\} = \mathbf{98\ mm}$ $p_2 = \min\{14t; 200\} = \min\{14 \times 7; 200\} = \mathbf{98\ mm}$ Try following spacing: e_1 = **40 mm** e_2 = **40 mm** p_1 = **60 mm** p_2 = **70 mm**	e_1 = 40 mm e_2 = 40 mm p_1 = 60 mm p_2 = 70 mm
7	Table 3.4	For bolt class 10.9, a_v = **0.5** For d = 24 mm, $A = \frac{\pi d^2}{4} = \frac{\pi \times 24^2}{4}$ $= \mathbf{452.40\ mm^2}$ Number of shear plane = Number of plate − 1 = 3 − 1	Number of bolt for shear resistance = 2

(continued)

(continued)

Step	Reference	Action/calculation	Conclusion
		$= \mathbf{2}$ Individual shear resistance per shear plane, $F_{v,Rd}$ $= \frac{a_v f_{ub} A}{\gamma_{M2}}$ $= \frac{0.5 \times 1000 \times 452.40}{1.25}$ $= 180.96$ kN *$F_{v,Rd}$ per bolt* *$=$ Individual $F_{v,Rd} \times$ shear plane* $= 180.96 \times 2$ $= \mathbf{361.92\ kN}$ *Number of bolt required* $= \frac{N_{Ed}}{F_{v,Rd}}$ $= \frac{500}{361.92}$ $= 1.4 = \mathbf{2}$	
8	Table 3.4	Conservatively, $a_b = \min\left\{\frac{e_1}{3d_0}; \frac{P_1}{3d_0} - \frac{1}{4}; \frac{f_{ub}}{f_u}; 1.0\right\}$ $= \min\left\{\frac{40}{3 \times 26}; \frac{60}{3 \times 26} - \frac{1}{4}; \frac{1000}{360}; 1.0\right\}$ $= \min\{0.51; 0.52; 2.78; 1.0\}$ $= \mathbf{0.51}$ $k_1 = \min\left\{2.8\frac{e_2}{d_0} - 1.7; 1.4\frac{P_2}{d_0} - 1.7; 2.5\right\}$ $= \min\left\{2.8 \times \frac{40}{26} - 1.7; 1.4 \times \frac{70}{26} - 1.7; 2.5\right\}$ $= \min\{2.6; 2.1; 2.5\}$ $= \mathbf{2.1}$ Individual bearing resistance, $F_{b,Rd}$ $= \frac{k_1 a_b f_u d t}{\gamma_{M2}}$ $= \frac{2.1 \times 0.51 \times 360 \times 24 \times 14}{1.25}$ $= \mathbf{103.64\ kN}$ *Number of bolt required* $= \frac{N_{Ed}}{F_{b,Rd}}$ $= \frac{500}{103.64}$ $= 4.8 = \mathbf{5}$	Number of bolt for bearing resistance = 5
9	Table 3.4	Individual tension resistance, $F_{t,Rd}$	

(continued)

(continued)

Step	Reference	Action/calculation	Conclusion
		$= \dfrac{k_2 f_{ub} A}{\gamma_{M2}}$ $= \dfrac{0.9 \times 1000 \times 452.40}{1.25}$ $= \mathbf{325.73\ kN}$ *Number of bolt required* $= \dfrac{N_{Ed}}{F_{t,Rd}}$ $= \dfrac{500}{325.73}$ $= 1.5 = \mathbf{2}$	Number of bolt for tensile resistance = 2
10		Number of bolt required = **5**	Number of bolt = 5
11		Check the following ratio: $\dfrac{N_{Ed}}{F_{v,Rd}} = \dfrac{500}{361.92 \times 5} = \mathbf{0.28}$ $\dfrac{N_{Ed}}{F_{b,Rd}} = \dfrac{500}{103.64 \times 5} = \mathbf{0.96}$ $\dfrac{N_{Ed}}{F_{t,Rd}} = \dfrac{500}{325.73 \times 5} = \mathbf{0.31}$ The bolts can be arranged in the way as shown below: 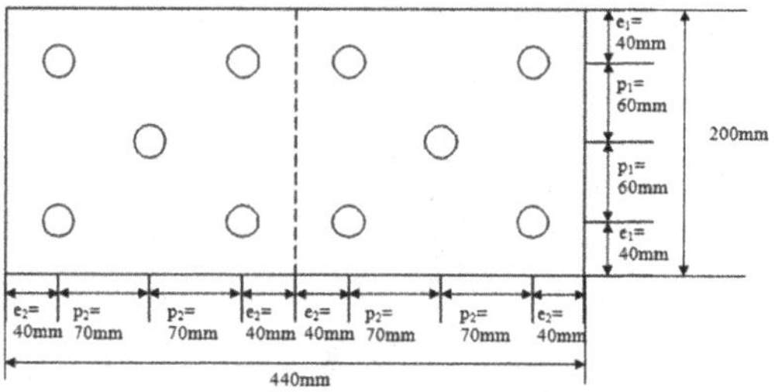 The minimum dimension of steel plate for such arrangement is 200 mm × 440 mm. Therefore, the steel plate suggested is **suitable** for this	

4.4 Exercise: Connection Design

4-1 Determine the minimum number of fillet welding sides required for the situation shown in Fig. 4.15. Steel grade S275 is used.

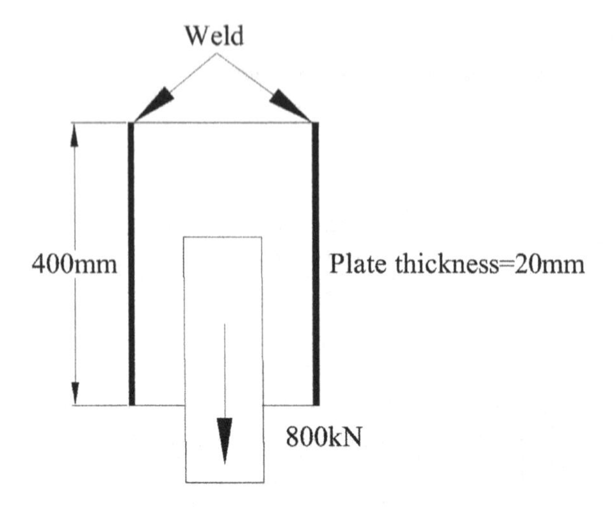

Fig. 4.15 Question 4-1

4-2 Determine the maximum resistance of the welded connection in the situation shown in Fig. 4.16. Steel grade S235 is used. The thickness of the steel plate is 15 mm.

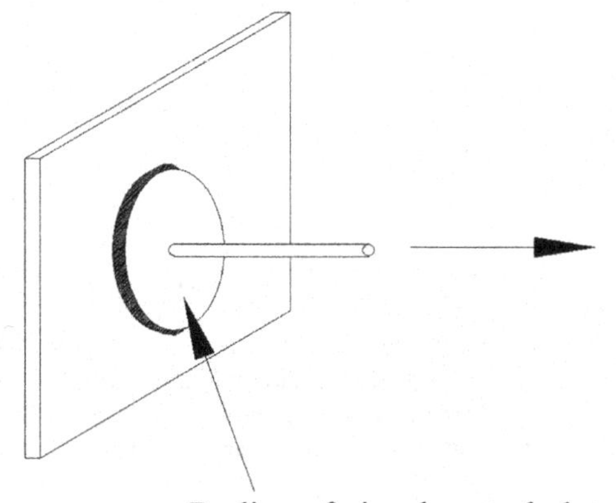

Fig. 4.16 Question 4-2

4-3 Determine the $\frac{N_{Ed}}{F_{v,Rd}}$, $\frac{N_{Ed}}{F_{b,Rd}}$, and $\frac{N_{Ed}}{F_{t,Rd}}$ ratios of the following bolted connection:

Design load	200 kN
Number of bolt	6
Bolt class	8.8
Diameter of bolt	20 mm

(continued)

(continued)

Design load	200 kN
Steel grade	S235
Number of steel plate	3
Plate thickness	8 mm each
e_1	30 mm
p_1	50 mm
e_2	30 mm
p_2	60 mm

4-4 Determine the minimum *size* of the steel plate required to establish both welded and bolted connections if the force of the bracing member is 750 kN, as shown in Fig. 4.17. Steel grade S275 is used.

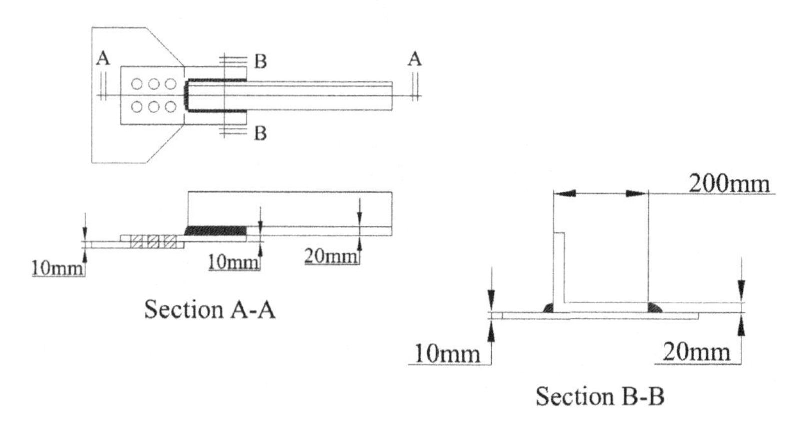

Fig. 4.17 Plan view and *size* view of connection, and section view of bracing member for Question 4-4

Appendix

See Tables A.1, A.2 and A.3.

F. Hejazi and T. K. Chun, *Steel Structures Design Based on Eurocode 3*,
https://doi.org/10.1007/978-981-10-8836-0

Table A.1 General formula for maximum shear, bending moment and deflection for several loading conditions

Loading condition	Reactions	Bending moment	Deflection
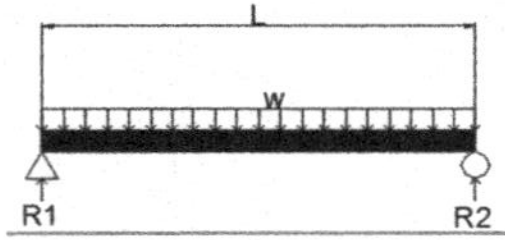 L, w, R1, R2	$R1 = R2 = \frac{wL}{2}$	$M_{\max} = \frac{wL^2}{8}$	$\Delta_{\max} = \frac{5wL^4}{384EI}$
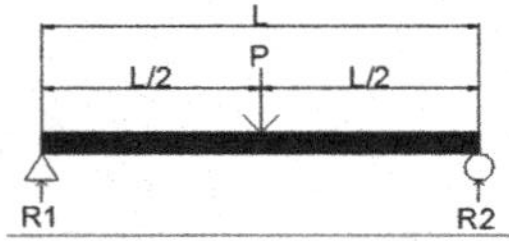 L, L/2, P, L/2, R1, R2	$R1 = R2 = \frac{P}{2}$	$M_{\max} = \frac{PL}{4}$	$\Delta_{\max} = \frac{PL^3}{48EI}$
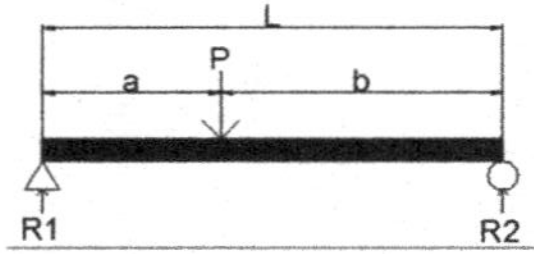 L, a, P, b, R1, R2	$R1 = \frac{Pb}{L}$ $R2 = \frac{Pa}{L}$	$M_{\max} = \frac{Pab}{L}$	$\Delta_{\max} = \frac{Pab(a+2b)\sqrt{3a(a+2b)}}{27EIL}$
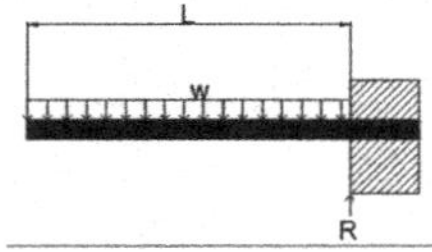 L, w, R	$R = wL$	$M_{\max} = \frac{wL^2}{2}$	$\Delta_{\max} = \frac{wL^4}{8EI}$
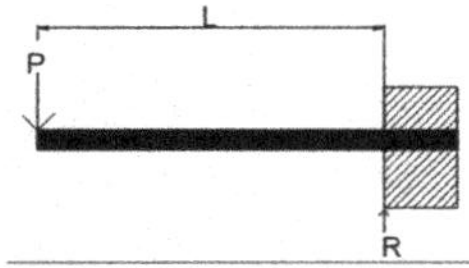 L, P, R	$R = P$	$M_{\max} = PL$	$\Delta_{\max} = \frac{PL^3}{3EI}$
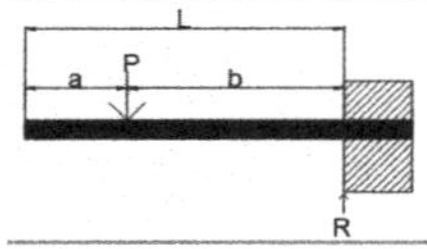 L, a, P, b, R	$R = P$	$M_{\max} = Pb$	$\Delta_{\max} = \frac{Pb^2}{6EI}(3L - b)$

(continued)

Table A.1 (continued)

Loading condition	Reactions	Bending moment	Deflection
	$R1 = \frac{3wL}{8}$ $R2 = \frac{5wL}{8}$	$M_{\max} = \frac{wL^2}{8}$	$\Delta_{\max} = \frac{wL^4}{185EI}$
	$R1 = \frac{5P}{16}$ $R2 = \frac{11P}{16}$	$M_{\max} = \frac{3PL}{16}$	$\Delta_{\max} = 0.009317\frac{PL^3}{EI}$
	$R1 = \frac{Pb^2}{2L^3}(a+2L)$ $R2 = \frac{Pa}{2L^3}(3L^2 - a^2)$	$M_1(\text{at point of load}) = R1a$ $M_2(\text{at fixed end})$ $= \frac{Pab}{2L^2}(a+L)$	$\Delta_{\max}(\textit{if } a < 0.414L)$ $= \frac{Pa}{3EI}\frac{(L^2 - a^2)^3}{(3L^2 - a^2)^2}$ $\Delta_{\max}(\textit{if } a > 0.414L)$ $= \frac{Pab^2}{6EI}\sqrt{\frac{a}{2L+a}}$
	$R1 = R2 = \frac{wL}{2}$	$M_{\max} = \frac{wL^2}{12}$	$\Delta_{\max} = \frac{wL^4}{384EI}$
	$R1 = R2 = \frac{P}{2}$	$M_{\max} = \frac{PL}{8}$	$\Delta_{\max} = \frac{PL^3}{192EI}$
	$R1 = \frac{Pb^2}{L^3}(3a+b)$ $R2 = \frac{Pa^3}{L^3}(a+3b)$	$M_1(\text{left end}) = \frac{Pab^2}{L^2}$ $M_2(\text{right end}) = \frac{Pa^2b}{L^2}$	$\Delta_{\max} = \frac{2Pa^3b^2}{3EI(3a+b)^2}$

Table A.2 Universal beam with sectional properties in EC notation (BS 4 Part 1 2005)

Designation	Mass per m	Depth of section	Width of section	Thickness		Root radius	Depth between fillets	Radius for local buckling		Second moment of area	
				Web	Flange			Flange	Web	Axis y-y	Axis z-z
		D	b	t_w	t_f	r	d	$b/2t_f$	d/t_w	I_y	I_z
	kg/m	mm	mm	mm	mm	mm	mm			cm^4	cm^4
127 × 76 × 13	13	127	76	4	7.6	7.6	96.6	5	24.1	473	55.7
152 × 89 × 16	16	152.4	88.7	4.5	7.7	7.6	121.8	5.76	27.1	834	89.8
178 × 102 × 19	19	177.8	101.2	4.8	7.9	7.6	146.8	6.41	30.6	1356	137
203 × 102 × 23	23.1	203.2	101.8	5.4	9.3	7.6	169.4	5.47	31.4	2105	164
203 × 133 × 25	25.1	203.2	133.2	5.7	7.8	7.6	172.4	8.54	30.2	2340	308
203 × 133 × 30	30	206.8	133.9	6.4	9.6	7.6	172.4	6.97	26.9	2896	385
254 × 102 × 22	22	254	101.6	5.7	6.8	7.6	225.2	7.47	39.5	2841	119
254 × 102 × 25	25.2	257.2	101.9	6	8.4	7.6	225.2	6.07	37.5	3415	149
254 × 102 × 28	28.3	260.4	102.2	6.3	10	7.6	225.2	5.11	35.7	4005	179
254 × 146 × 31	31.1	251.4	146.1	6	8.6	7.6	219	8.49	36.5	4413	448
254 × 146 × 37	37	256	146.4	6.3	10.9	7.6	219	6.72	34.8	5537	571
254 × 146 × 43	43	259.6	147.3	7.2	12.7	7.6	219	5.8	30.4	6544	677
305 × 102 × 25	24.8	305.1	101.6	5.8	7	7.6	275.9	7.26	47.6	4455	123
305 × 102 × 28	28.2	308.7	101.8	6	8.8	7.6	275.9	5.78	46	5366	155
305 × 102 × 33	32.8	312.7	102.4	6.6	10.8	7.6	275.9	4.74	41.8	6501	194
305 × 127 × 37	37	304.4	123.4	7.1	10.7	8.9	265.2	5.77	37.4	7171	336
305 × 127 × 42	41.9	307.2	124.3	8	12.1	8.9	265.2	5.14	33.1	8196	389
305 × 127 × 48	48.1	311	125.3	9	14	8.9	265.2	4.47	29.5	9575	461

(continued)

Table A.2 (continued)

Designation	Mass per m	Depth of section	Width of section	Thickness		Root radius	Depth between fillets	Radius for local buckling		Second moment of area	
				Web	Flange			Flange	Web	Axis y-y	Axis z-z
		D	b	t_w	t_f	r	d	$b/2t_f$	d/t_w	I_y	I_z
	kg/m	mm	mm	mm	mm	mm	mm			cm^4	cm^4
305 × 165 × 40	40.3	303.4	165	6	10.2	8.9	265.2	8.09	44.2	8503	764
305 × 165 × 46	46.1	306.6	165.7	6.7	11.8	8.9	265.2	7.02	39.6	9899	896
305 × 165 × 54	54	310.4	166.9	7.9	13.7	8.9	265.2	6.09	33.6	11,700	1063
356 × 127 × 33	33.1	349	125.4	6	8.5	10.2	311.6	7.38	51.9	8249	280
356 × 127 × 39	39.1	353.4	126	6.6	10.7	10.2	311.6	5.89	47.2	10,170	358
356 × 171 × 45	45	351.4	171.1	7	9.7	10.2	311.6	8.82	44.5	12,070	811
356 × 171 × 51	51	355	171.5	7.4	11.5	10.2	311.6	7.46	42.1	14,140	968
356 × 171 × 57	57	358	172.2	8.1	13	10.2	311.6	6.62	38.5	16,040	1108
356 × 171 × 67	67.1	363.4	173.2	9.1	15.7	10.2	311.6	5.52	34.2	19,460	1362
406 × 140 × 39	39	398	141.8	6.4	8.6	10.2	360.4	8.24	56.3	12,510	410
406 × 140 × 46	46	403.2	142.2	6.8	11.2	10.2	360.4	6.35	53	15,690	538
406 × 178 × 54	54.1	402.6	177.7	7.7	10.9	10.2	360.4	8.15	46.8	18,720	1021
406 × 178 × 60	60.1	406.4	177.9	7.9	12.8	10.2	360.4	6.95	45.6	21,600	1203
406 × 178 × 67	67.1	409.4	178.8	8.8	14.3	10.2	360.4	6.25	41	24,330	1365
406 × 178 × 74	74.2	412.8	179.5	9.5	16	10.2	360.4	5.61	37.9	27,310	1545
457 × 152 × 52	52.3	449.8	152.4	7.6	10.9	10.2	407.6	6.99	53.6	21,370	645
457 × 152 × 60	59.8	454.6	152.9	8.1	13.3	10.2	407.6	5.75	50.3	25,500	795
457 × 152 × 67	67.2	458	153.8	9	15	10.2	407.6	5.13	45.3	28,930	913

(continued)

Table A.2 (continued)

Designation	Mass per m	Depth of section	Width of section	Thickness		Root radius	Depth between fillets	Radius for local buckling		Second moment of area	
				Web	Flange			Flange	Web	Axis y-y	Axis z-z
		D	b	t_w	t_f	r	d	$b/2t_f$	d/t_w	I_y	I_z
	kg/m	mm	mm	mm	mm	mm	mm			cm^4	cm^4
457 × 152 × 74	74.2	462	154.4	9.6	17	10.2	407.6	4.54	42.5	32,670	1047
457 × 152 × 82	82.1	465.8	155.3	10.5	18.9	10.2	407.6	4.11	38.8	36,590	1185
457 × 191 × 67	67.1	453.4	189.9	8.5	12.7	10.2	407.6	7.48	48	29,380	1452
457 × 191 × 74	74.3	457	190.4	9	14.5	10.2	407.6	6.57	45.3	33,320	1671
457 × 191 × 82	82	460	191.3	9.9	16	10.2	407.6	5.98	41.2	37,050	1871
457 × 191 × 89	89.3	463.4	191.9	10.5	17.7	10.2	407.6	5.42	38.8	41,020	2089
457 × 191 × 98	98.3	467.2	192.8	11.4	19.6	10.2	407.6	4.92	35.8	45,730	2347
533 × 210 × 101	101	536.7	210	10.8	17.4	12.7	476.5	6.03	44.1	61,520	2692
533 × 210 × 109	109	539.5	210.8	11.6	18.8	12.7	476.5	5.61	41.1	66,820	2943
533 × 210 × 122	122	544.5	211.9	12.7	21.3	12.7	476.5	4.97	37.5	76,040	3388
533 × 210 × 82	82.2	528.3	208.8	9.6	13.2	12.7	476.5	7.91	49.6	47,540	2007
533 × 210 × 92	92.14	533.1	209.3	10.1	15.6	12.7	476.5	6.71	47.2	55,230	2389
610 × 229 × 101	101.2	602.6	227.6	10.5	14.8	12.7	547.6	7.69	52.2	75,780	2915
610 × 229 × 113	113	607.6	228.2	11.1	17.3	12.7	547.6	6.6	49.3	87,320	3434
610 × 229 × 125	125.1	612.2	229	11.9	19.6	12.7	547.6	5.84	46	98,610	3932
610 × 229 × 140	139.9	617.2	230.2	13.1	22.1	12.7	547.6	5.21	41.8	111,800	4505
610 × 305 × 149	149.2	612.4	304.8	11.8	19.7	16.5	540	7.74	45.8	125,900	9308
610 × 305 × 179	179	620.2	307.1	14.1	23.6	16.5	540	6.51	38.3	153,000	11,410

(continued)

Table A.2 (continued)

Designation	Mass per m	Depth of section	Width of section	Thickness		Root radius	Depth between fillets	Radius for local buckling		Second moment of area	
				Web	Flange			Flange	Web	Axis *y-y*	Axis *z-z*
		D	b	t_w	t_f	r	d	$b/2t_f$	d/t_w	I_y	I_z
	kg/m	mm	mm	mm	mm	mm	mm			cm^4	cm^4
610 × 305 × 238	238.1	635.8	311.4	18.4	31.4	16.5	540	4.96	29.3	209,500	15,840
686 × 254 × 125	125.2	677.9	253	11.7	16.2	15.2	615.1	7.81	52.6	118,000	4383
686 × 254 × 140	140.1	683.5	253.7	12.4	19	15.2	615.1	6.68	49.6	136,300	5183
686 × 254 × 152	152.4	687.5	254.5	13.2	21	15.2	615.1	6.06	46.6	150,400	5784
686 × 254 × 170	170.2	692.9	255.8	14.5	23.7	15.2	615.1	5.4	42.4	170,300	6630
762 × 267 × 134	133.9	750	264.4	12	15.5	16.5	686	8.53	57.2	150,700	4788
762 × 267 × 147	146.9	754	265.2	12.8	17.5	16.5	686	7.58	53.6	168,500	5455
762 × 267 × 173	173	762.2	266.7	14.3	21.6	16.5	686	6.17	48	205,300	6850
762 × 267 × 197	196.8	769.8	268	15.6	25.4	16.5	686	5.28	44	240,000	8175
838 × 292 × 176	175.9	834.9	291.7	14	18.8	17.8	761.7	7.76	54.4	246,000	7799
838 × 292 × 194	193.8	840.7	292.4	14.7	21.7	17.8	761.7	6.74	51.8	279,200	9066
838 × 292 × 226	226.5	850.9	293.8	16.1	26.8	17.8	761.7	5.48	47.3	339,700	11,360
914 × 305 × 201	200.9	903	303.3	15.1	20.2	19.1	824.4	7.51	54.6	325,300	9423
914 × 305 × 224	224.2	910.4	304.1	15.9	23.9	19.1	824.4	6.36	51.8	376,400	11,240
914 × 305 × 253	253.4	918.4	305.5	17.3	27.9	19.1	824.4	5.47	47.7	436,300	13,300
914 × 305 × 289	289.1	926.6	307.7	19.5	32	19.1	824.4	4.81	42.3	504,200	15,600
914 × 419 × 343	343.3	911.8	418.5	19.4	32	24.1	799.6	6.54	41.2	625,800	39,160
914 × 419 × 388	388	921	420.5	21.4	36.6	24.1	799.6	5.74	37.4	719,600	45,440

Table A.2 (continued)

Designation	Radius of gyration		Elastic modulus		Plastic modulus		Buckling parameter	Torsional index	Warping constant	Torsional constant	Area of section
	Axis y-y	Axis z-z	Axis y-y	Axis z-z	Axis y-y	Axis z-z					
	i_y	i_z	$W_{el,y}$	$W_{el,z}$	$W_{pl,y}$	$W_{pl,z}$	u	x	I_w	I_t	A
	cm	cm	cm^3	cm^3	cm^3	cm^3			dm^6	cm^4	cm^2
127 × 76 × 13	5.35	1.84	74.6	14.7	84.2	22.6	0.895	16.3	0.002	2.85	16.5
152 × 89 × 16	6.41	2.1	109	20.2	123	31.2	0.89	19.6	0.005	3.56	20.3
178 × 102 × 19	7.48	2.37	153	27	171	41.6	0.888	22.6	0.01	4.41	24.3
203 × 102 × 23	8.46	2.36	207	32.2	234	49.8	0.888	22.5	0.015	7.02	29.4
203 × 133 × 25	8.56	3.1	230	46.2	258	70.9	0.877	25.6	0.029	5.96	32
203 × 133 × 30	8.71	3.17	280	57.5	314	88.2	0.881	21.5	0.037	10.3	38.2
254 × 102 × 22	10.1	2.06	224	23.5	259	37.3	0.856	36.4	0.018	4.15	28
254 × 102 × 25	10.3	2.15	266	29.2	306	46	0.866	31.5	0.023	6.42	32
254 × 102 × 28	10.5	2.22	308	34.9	353	54.8	0.874	27.5	0.028	9.57	36.1
254 × 146 × 31	10.5	3.36	351	61.3	393	94.1	0.88	29.6	0.066	8.55	39.7
254 × 146 × 37	10.8	3.48	433	78	483	119	0.89	24.3	0.086	15.3	47.2
254 × 146 × 43	10.9	3.52	504	92	566	141	0.891	21.2	0.103	23.9	54.8
305 × 102 × 25	11.9	1.97	292	24.2	342	38.8	0.846	43.4	0.027	4.77	31.6
305 × 102 × 28	12.2	2.08	348	30.5	403	48.5	0.859	37.4	0.035	7.4	35.9
305 × 102 × 33	12.5	2.15	416	37.9	481	60	0.866	31.6	0.044	12.2	41.8
305 × 127 × 37	12.3	2.67	471	54.5	539	85.4	0.872	29.7	0.072	14.8	47.2
305 × 127 × 42	12.4	2.7	534	62.6	614	98.4	0.872	26.5	0.085	21.1	53.4
305 × 127 × 48	12.5	2.74	616	73.6	711	116	0.873	23.3	0.102	31.8	61.2
305 × 165 × 40	12.9	3.86	560	92.6	623	142	0.889	31	0.164	14.7	51.3

(continued)

Table A.2 (continued)

Designation	Radius of gyration		Elastic modulus		Plastic modulus		Buckling parameter	Torsional index	Warping constant	Torsional constant	Area of section
	Axis y-y	Axis z-z	Axis y-y	Axis z-z	Axis y-y	Axis z-z					
	i_y	i_z	$W_{el,y}$	$W_{el,z}$	$W_{pl,y}$	$W_{pl,z}$	u	x	I_w	I_t	A
	cm	cm	cm^3	cm^3	cm^3	cm^3			dm^6	cm^4	cm^2
305 × 165 × 46	13	3.9	646	108	720	166	0.891	27.1	0.195	22.2	58.7
305 × 165 × 54	13	3.93	754	127	846	196	0.889	23.6	0.234	34.8	68.8
356 × 127 × 33	14	2.58	473	44.7	543	70.3	0.863	42.2	0.081	8.79	42.1
356 × 127 × 39	14.3	2.68	576	56.8	659	89.1	0.871	35.2	0.105	15.1	49.8
356 × 171 × 45	14.5	3.76	687	94.8	775	147	0.874	36.8	0.237	15.8	57.3
356 × 171 × 51	14.8	3.86	796	113	896	174	0.881	32.1	0.286	23.8	64.9
356 × 171 × 57	14.9	3.91	896	129	1010	199	0.882	28.8	0.33	33.4	72.6
356 × 171 × 67	15.1	3.99	1071	157	1211	243	0.886	24.4	0.412	55.7	85.5
406 × 140 × 39	15.9	2.87	629	57.8	724	90.8	0.858	47.5	0.155	10.7	49.7
406 × 140 × 46	16.4	3.03	778	75.7	888	118	0.871	38.9	0.207	19	58.6
406 × 178 × 54	16.5	3.85	930	115	1055	178	0.871	38.3	0.392	23.1	69
406 × 178 × 60	16.8	3.97	1063	135	1199	209	0.88	33.8	0.466	33.3	76.5
406 × 178 × 67	16.9	3.99	1189	153	1346	237	0.88	30.5	0.533	46.1	85.5
406 × 178 × 74	17	4.04	1323	172	1501	267	0.882	27.6	0.608	62.8	94.5
457 × 152 × 52	17.9	3.11	950	84.6	1096	133	0.859	43.9	0.311	21.4	66.6
457 × 152 × 60	18.3	3.23	1122	104	1287	163	0.868	37.5	0.387	33.8	76.2
457 × 152 × 67	18.4	3.27	1263	119	1453	187	0.869	33.6	0.448	47.7	85.6
457 × 152 × 74	18.6	3.33	1414	136	1627	213	0.873	30.1	0.518	65.9	94.5
457 × 152 × 82	18.7	3.37	1571	153	1811	240	0.873	27.4	0.591	89.2	105

(continued)

Table A.2 (continued)

Designation	Radius of gyration		Elastic modulus		Plastic modulus		Buckling parameter	Torsional index	Warping constant	Torsional constant	Area of section
	Axis y-y	Axis z-z	Axis y-y	Axis z-z	Axis y-y	Axis z-z					
	i_y	i_z	$W_{el,y}$	$W_{el,z}$	$W_{pl,y}$	$W_{pl,z}$	u	x	I_w	I_t	A
	cm	cm	cm^3	cm^3	cm^3	cm^3			dm^6	cm^4	cm^2
457 × 191 × 67	18.5	4.12	1296	153	1471	237	0.872	37.9	0.705	37.1	85.5
457 × 191 × 74	18.8	4.2	1458	176	1653	272	0.877	33.9	0.818	51.8	94.6
457 × 191 × 82	18.8	4.23	1611	196	1831	304	0.877	30.9	0.922	69.2	104
457 × 191 × 89	19	4.29	1770	218	2014	338	0.88	28.3	1.04	90.7	114
457 × 191 × 98	19.1	4.33	1957	243	2232	379	0.881	25.7	1.18	121	125
533 × 210 × 101	21.9	4.57	2292	256	2612	399	0.874	33.2	1.81	101	129
533 × 210 × 109	21.9	4.6	2477	279	2828	436	0.875	30.9	1.99	126	139
533 × 210 × 122	22.1	4.67	2793	320	3196	500	0.877	27.6	2.32	178	155
533 × 210 × 82	21.3	4.38	1800	192	2059	300	0.864	41.6	1.33	51.5	105
533 × 210 × 92	21.7	4.51	2072	228	2360	356	0.872	36.5	1.6	75.7	117
610 × 229 × 101	24.2	4.75	2515	256	2881	400	0.864	43.1	2.52	77	129
610 × 229 × 113	24.6	4.88	2874	301	3281	469	0.87	38	2.99	111	144
610 × 229 × 125	24.9	4.97	3221	343	3676	535	0.873	34.1	3.45	154	159
610 × 229 × 140	25	5.03	3622	391	4142	611	0.875	30.6	3.99	216	178
610 × 305 × 149	25.7	7	4111	611	4594	937	0.886	32.7	8.17	200	190
610 × 305 × 179	25.9	7.07	4935	743	5547	1144	0.886	27.7	10.2	340	228
610 × 305 × 238	26.3	7.23	6589	1017	7486	1574	0.886	21.3	14.5	785	303
686 × 254 × 125	27.2	5.24	3481	346	3994	542	0.862	43.9	4.8	116	159
686 × 254 × 140	27.6	5.39	3987	409	4558	638	0.868	38.7	5.72	169	178

(continued)

Table A.2 (continued)

Designation	Radius of gyration		Elastic modulus		Plastic modulus		Buckling parameter	Torsional index	Warping constant	Torsional constant	Area of section
	Axis y-y	Axis z-z	Axis y-y	Axis z-z	Axis y-y	Axis z-z					
	i_y	i_z	$W_{el,y}$	$W_{el,z}$	$W_{pl,y}$	$W_{pl,z}$	u	x	I_w	I_t	A
	cm	cm	cm^3	cm^3	cm^3	cm^3			dm^6	cm^4	cm^2
686 × 254 × 152	27.8	5.46	4374	455	5000	710	0.871	35.5	6.42	220	194
686 × 254 × 170	28	5.53	4916	518	5631	811	0.872	31.8	7.42	308	217
762 × 267 × 134	29.7	5.3	4018	362	4644	570	0.854	49.8	6.46	119	171
762 × 267 × 147	30	5.4	4470	411	5156	647	0.858	45.2	7.4	159	187
762 × 267 × 173	30.5	5.58	5387	514	6198	807	0.864	38.1	9.39	267	220
762 × 267 × 197	30.9	5.71	6234	610	7167	959	0.869	33.2	11.3	404	251
838 × 292 × 176	33.1	5.9	5893	535	6808	842	0.856	46.5	13	221	224
838 × 292 × 194	33.6	6.06	6641	620	7640	974	0.862	41.6	15.2	306	247
838 × 292 × 226	34.3	6.27	7985	773	9155	1212	0.87	35	19.3	514	289
914 × 305 × 201	35.7	6.07	7204	621	8351	982	0.854	46.8	18.4	291	256
914 × 305 × 224	36.3	6.27	8269	739	9535	1163	0.861	41.3	22.1	422	286
914 × 305 × 253	36.8	6.42	9501	871	10,940	1371	0.866	36.2	26.4	626	323
914 × 305 × 289	37	6.51	10,880	1014	12,570	1601	0.867	31.9	31.2	926	368
914 × 419 × 343	37.8	9.46	13,730	1871	15,480	2890	0.883	30.1	75.8	1193	437
914 × 419 × 388	38.2	9.59	15,630	2161	17,670	3341	0.885	26.7	88.9	1734	494

Table A.3 Universal column with sectional properties in EC notation (BS 4 Part 1 2005)

Designation	Mass per m	Depth of section	Width of section	Thickness		Root radius	Depth between fillets	Radius for local buckling		Second moment of area	
				Web	Flange			Flange	Web	Axis y-y	Axis z-z
		D	b	t_w	t_f	r	d	$b/2t_f$	d/t_w	I_y	I_z
	kg/m	mm	mm	mm	mm	mm	mm			cm^4	cm^4
152 × 152 × 23	23	152.4	152.2	5.8	6.8	7.6	123.6	11.2	21.3	1250	400
152 × 152 × 30	30	157.6	152.9	6.5	9.4	7.6	123.6	8.13	19	1748	560
152 × 152 × 37	37	161.8	154.4	8	11.5	7.6	123.6	6.71	15.5	2210	706
203 × 203 × 46	46.1	203.2	203.6	7.2	11	10.2	160.8	9.25	22.3	4568	1548
203 × 203 × 52	52	206.2	204.3	7.9	12.5	10.2	160.8	8.17	20.4	5259	1778
203 × 203 × 60	60	209.6	205.8	9.4	14.2	10.2	160.8	7.25	17.1	6125	2065
203 × 203 × 71	71	215.8	206.4	10	17.3	10.2	160.8	5.97	16.1	7618	2537
203 × 203 × 86	86.1	222.2	209.1	12.7	20.5	10.2	160.8	5.1	12.7	9449	3127
254 × 254 × 107	107.1	266.7	258.8	12.8	20.5	12.7	200.3	6.31	15.6	17,510	5928
254 × 254 × 132	132	276.3	261.3	15.3	25.3	12.7	200.3	5.16	13.1	22,530	7531
254 × 254 × 167	167.1	289.1	265.2	19.2	31.7	12.7	200.3	4.18	10.4	30,000	9870
254 × 254 × 73	73.1	254.1	254.6	8.6	14.2	12.7	200.3	8.96	23.3	11,410	3908
254 × 254 × 89	88.9	260.3	256.3	10.3	17.3	12.7	200.3	7.41	19.4	14,270	4857
305 × 305 × 118	117.9	314.5	307.4	12	18.7	15.2	246.7	8.22	20.6	27,670	9059
305 × 305 × 137	136.9	320.5	309.2	13.8	21.7	15.2	246.7	7.12	17.9	32,810	10,700
305 × 305 × 158	158.1	327.1	311.2	15.8	25	15.2	246.7	6.22	15.6	38,750	12,570
305 × 305 × 198	198.1	339.9	314.5	19.1	31.4	15.2	246.7	5.01	12.9	50,900	16,300
305 × 305 × 240	240	352.5	318.4	23	37.7	15.2	246.7	4.22	10.7	64,200	20,310
305 × 305 × 283	282.9	365.3	322.2	26.8	44.1	15.2	246.7	3.65	9.21	78,870	24,630
305 × 305 × 97	96.9	307.9	305.3	9.9	15.4	15.2	246.7	9.91	24.9	22,250	7308
356 × 368 × 129	129	355.6	368.6	10.4	17.5	15.2	290.2	10.5	27.9	40,250	14,610

(continued)

Table A.3 (continued)

Designation	Mass per m	Depth of section	Width of section	Thickness		Root radius	Depth between fillets	Radius for local buckling		Second moment of area	
				Web	Flange			Flange	Web	Axis y-y	Axis z-z
		D	b	t_w	t_f	r	d	$b/2t_f$	d/t_w	I_y	I_z
	kg/m	mm	mm	mm	mm	mm	mm			cm^4	cm^4
356 × 368 × 153	152.9	362	370.5	12.3	20.7	15.2	290.2	8.95	23.6	48,590	17,550
356 × 368 × 177	177	368.2	372.6	14.4	23.8	15.2	290.2	7.83	20.2	57,120	20,530
356 × 368 × 202	201.9	374.6	374.7	16.5	27	15.2	290.2	6.94	17.6	66,260	23,690
356 × 406 × 235	235.1	381	394.8	18.4	30.2	15.2	290.2	6.54	15.8	79,080	30,990
356 × 406 × 287	287.1	393.6	399	22.6	36.5	15.2	290.2	5.47	12.8	99,880	38,680
356 × 406 × 340	339.9	406.4	403	26.6	42.9	15.2	290.2	4.7	10.9	122,500	46,850
356 × 406 × 393	393	419	407	30.6	49.2	15.2	290.2	4.14	9.48	146,600	55,370
356 × 406 × 467	467	436.6	412.2	35.8	58	15.2	290.2	3.55	8.11	183,000	67,830
356 × 406 × 551	551	455.6	418.5	42.1	67.5	15.2	290.2	3.1	6.89	226,900	82,670
356 × 406 × 634	633.9	474.6	424	47.6	77	15.2	290.2	2.75	6.1	274,800	98,130

Designation	Radius of gyration		Elastic modulus		Plastic modulus		Buckling parameter	Torsional Index	Warping constant	Torsional constant	Area of section
	Axis y-y	Axis z-z	Axis y-y	Axis z-z	Axis y-y	Axis z-z					
	i_y	i_z	$W_{el,y}$	$W_{el,z}$	$W_{pl,y}$	$W_{pl,z}$	u	x	I_w	I_t	A
	cm	cm	cm^3	cm^3	cm^3	cm^3			dm^6	cm^4	cm^2
152 × 152 × 23	6.54	3.7	164	52.6	182	80.2	0.84	20.7	0.021	4.63	29.2
152 × 152 × 30	6.76	3.83	222	73.3	248	112	0.849	16	0.031	10.5	38.3
152 × 152 × 37	6.85	3.87	273	91.5	309	140	0.848	13.3	0.04	19.2	47.1
203 × 203 × 46	8.82	5.13	450	152	497	231	0.847	17.7	0.143	22.2	58.7
203 × 203 × 52	8.91	5.18	510	174	567	264	0.848	15.8	0.167	31.8	66.3

(continued)

Table A.3 (continued)

Designation	Radius of gyration		Elastic modulus		Plastic modulus		Buckling parameter	Torsional Index	Warping constant	Torsional constant	Area of section
	Axis *y-y*	Axis *z-z*	Axis *y-y*	Axis *z-z*	Axis *y-y*	Axis *z-z*					
	i_y	i_z	$W_{el,y}$	$W_{el,z}$	$W_{pl,y}$	$W_{pl,z}$	u	x	I_w	I_t	A
	cm	cm	cm^3	cm^3	cm^3	cm^3			dm^6	cm^4	cm^2
203 × 203 × 60	8.96	5.2	584	201	656	305	0.846	14.1	0.197	47.2	76.4
203 × 203 × 71	9.18	5.3	706	246	799	374	0.853	11.9	0.25	80.2	90.4
203 × 203 × 86	9.28	5.34	850	299	977	456	0.85	10.2	0.318	137	110
254 × 254 × 107	11.3	6.59	1313	458	1484	697	0.848	12.4	0.898	172	136
254 × 254 × 132	11.6	6.69	1631	576	1869	878	0.85	10.3	1.19	319	168
254 × 254 × 167	11.9	6.81	2075	744	2424	1137	0.851	8.49	1.63	626	213
254 × 254 × 73	11.1	6.48	898	307	992	465	0.849	17.3	0.562	57.6	93.1
254 × 254 × 89	11.2	6.55	1096	379	1224	575	0.85	14.5	0.717	102	113
305 × 305 × 118	13.6	7.77	1760	589	1958	895	0.85	16.2	1.98	161	150
305 × 305 × 137	13.7	7.83	2048	692	2297	1053	0.851	14.2	2.39	249	174
305 × 305 × 158	13.9	7.9	2369	808	2680	1230	0.851	12.5	2.87	378	201
305 × 305 × 198	14.2	8.04	2995	1037	3440	1581	0.854	10.2	3.88	734	252
305 × 305 × 240	14.5	8.15	3643	1276	4247	1951	0.854	8.74	5.03	1271	306
305 × 305 × 283	14.8	8.27	4318	1529	5105	2342	0.855	7.65	6.35	2034	360
305 × 305 × 97	13.4	7.69	1445	479	1592	726	0.85	19.3	1.56	91.2	123
356 × 368 × 129	15.6	9.43	2264	793	2479	1199	0.844	19.9	4.18	153	164
356 × 368 × 153	15.8	9.49	2684	948	2965	1435	0.844	17	5.11	251	195
356 × 368 × 177	15.9	9.54	3103	1102	3455	1671	0.844	15	6.09	381	226
356 × 368 × 202	16.1	9.6	3538	1264	3972	1920	0.844	13.4	7.16	558	257

(continued)

Table A.3 (continued)

Designation	Radius of gyration		Elastic modulus		Plastic modulus		Buckling parameter	Torsional Index	Warping constant	Torsional constant	Area of section
	Axis y-y	Axis z-z	Axis y-y	Axis z-z	Axis y-y	Axis z-z					
	i_y	i_z	$W_{el,y}$	$W_{el,z}$	$W_{pl,y}$	$W_{pl,z}$	u	x	I_w	I_t	A
	cm	cm	cm^3	cm^3	cm^3	cm^3			dm^6	cm^4	cm^2
356 × 406 × 235	16.3	10.2	4151	1570	4687	2383	0.834	12.1	9.54	812	299
356 × 406 × 287	16.5	10.3	5075	1939	5812	2949	0.835	10.2	12.3	1441	366
356 × 406 × 340	16.8	10.4	6031	2325	6999	3544	0.836	8.85	15.5	2343	433
356 × 406 × 393	17.1	10.5	6998	2721	8222	4154	0.837	7.86	18.9	3545	501
356 × 406 × 467	17.5	10.7	8383	3291	10,000	5034	0.839	6.86	24.3	5809	595
356 × 406 × 551	18	10.9	9962	3951	12,080	6058	0.841	6.05	31.1	9240	702
356 × 406 × 634	18.4	11	11,580	4629	14,240	7108	0.843	5.46	38.8	13,720	808

References

British Standard Institution. (2000). *BS 5950: Structural use of steelwork in building—part 1: Code of practice for design—rolled and welded sections*. London.

British Standard Institution. (2004). *BS EN 10025 Hot rolled products of structural steels—part 2: Technical delivery conditions for non-alloy structural steels*. London.

British Standard Institution. (2005a). *BS EN 1993 Eurocode 3: Design of steel structures—part 1–1: General rules and rules for buildings*. London.

British Standard Institution. (2005b). *BS EN 1993 Eurocode 3: Design of steel structures—part 1–8: Design of joints*. London.

British Standard Institution. (2005c). *NA to BS EN 1990UK National Annex for Eurocode: Basis of structural design*. London.

British Standard Institution. (2005d). *BS 4Structural steel sections—part 1: Specification for hot-rolled sections*. London.

British Standard Institution. (2008). *NA to BS EN 1993 UK National Annex to Eurocode 3: Design of steel structures—part 1-1: General rules and rules for buildings*. London.

NCCI: Elastic critical moment for lateral torsional buckling. SN003b.doc. Access Steel.

NCCI: Determination of moments on columns in simple construction. SN005a-EN-EU.doc. Access Steel.

NCCI: Verification of columns in simple construction—a simplified interaction criterion. SN048b-EN-GB.doc. Access Steel.

Further Reading

Arya, C. (2009). *Design of structural elements: Concrete, steelwork, masonry and timber design to British Standards and Eurocodes* (3rd ed.). New York: Taylor & Francis.

Brettle, M. E., & Brown, D. G. (Eds.). (2009). *Steel building design: Worked examples for students in accordance with Eurocodes and the UK National Annexes*. Berkshire: The Steel Construction Institution.

Gardner, L., & Nethercot, D. (2011). *Designer's guide to Eurocodes 3: Design of steel buildings EN 1993-1-1, -1-3 and -1-8* (2nd ed.). London: ICE Publishing.

F. Hejazi and T. K. Chun, *Steel Structures Design Based on Eurocode 3*,
https://doi.org/10.1007/978-981-10-8836-0

Printed in the United States
By Bookmasters